Renewable Energy

By Mark Boxall

2023 Edition

Copyright © 2023 by Mark Boxall

<u>Sparkyb Productions</u>

All rights reserved. No part of this publication may be reproduced, stored or transmitted in any form or by any means, electronic, mechanical, photocopying, recording, scanning, or otherwise without written permission from the publisher. It is illegal to copy this book, post it to a website, or distribute it by any other means without permission.

Mark Boxall asserts the moral right to be identified as the author of this work.

Mark Boxall has no responsibility for the persistence or accuracy of URLs for external or third-party Internet Websites referred to in this publication and does not guarantee that any content on such Websites is, or will remain, accurate or appropriate.

Designations used by companies to distinguish their products are often claimed as trademarks. All brand names and product names used in this book and on its cover are trade names, service marks, trademarks and registered trademarks of their respective owners. The publishers and the book are not associated with any product or vendor mentioned in this book. None of the companies referenced within the book have endorsed the book.

2023 Edition

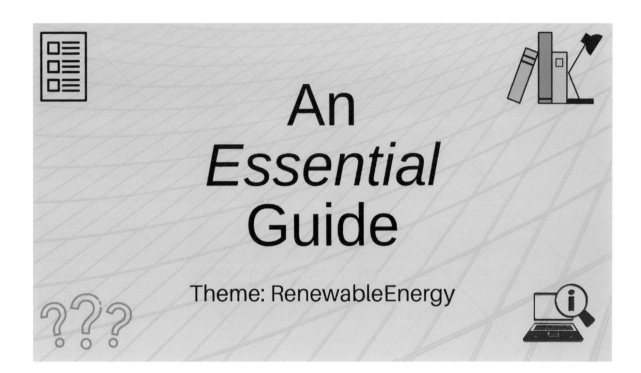

Knowledge is power. By increasing your knowledge you increase your ability to influence people and change attitudes.

Essential guides are designed to take advantage of the e-book format by providing you with all the tools you need to understand the basics of the subject, access the latest information and study any aspect in as much details as you want.

The basics are covered in an easy to understand format with no jargon and all relevant technical terms explained. There are clear diagrams, historical facts and much more. Also included at the end of each chapter are recommended further reading links to websites that contain videos, articles and more detailed explanations so you can expand your knowledge as much as you want, get the latest information and even find work or investment opportunities. How much you get out of it is up to you.

Using an Essential Guide will save you time and frustration. No more decisions about which search engine is best, no more frustration with broken links, out of date sites or irrelevant information. We have done all of that for you and included links to every site that we used in the process in the reference section.

Obviously there are no embedded links in the print editions, you will need to type any URLs yourself.

https://sparkybproductions.mysites.io/home

Dedication

My Wife

Arlene is a comfort and support in difficult times and I love her very much

My Family

My father David is no longer with us and my mother Diana is doing her best to adapt to life without him but still finds the time to help me with my writing. Her input and encouragement is a constant driving force, as is my daughter Sara.

Extended Family

Jason, Thomas, Suzie & Jay, Geoffrey, all of whom have supported me in times of need.

Friends

Paul and Ryan

Contents

INTRODUCTION 7

ELECTRICITY 10

What is electricity? 10

Electricity in Human History 12

Using Electricity 17

How Electricity is produced, transmitted and stored. 18
Electricity transmission and distribution 18

Batteries 20
Major developments in battery technology 21
The impact of batteries on the environment 22

H IS FOR HYDROGEN 24

What is hydrogen? 24

Hydrogen in human history 28
17th Century – The Age of Reason 28
18th Century – The Age of Enlightenment 29
19th Century 29
Airships 32
20th Century 32
Nuclear Weapons 33
Post War 34
21st Century 35

How hydrogen is produced, processed and stored 37
Hydrogen production methods 38

Hydrogen Storage Methods 45

Practical application and uses of hydrogen 48
Overview 48
Industrial use 48
Fuel Cells 51

Where are we now? 62
Generating or manufacturing hydrogen fuel 62
Hydrogen fuel vehicles 63
Other vehicles 64
Static Generation 71

OTHER CLEAN ENERGY PRODUCTION METHODS 73

Solar Energy 73
Concentrated Solar Power 74
Photovoltaic (PV) 75

Hydro Power 78

Energy from our seas and oceans 80

Geothermal 82

Biomass 83

Nuclear Power 85

Wind Power 87

ADVANTAGES AND DISADVANTAGES 89

Large scale energy production (for power grids) 89

Small scale energy production (for local and individual use) 92

Combining different technologies 94

NEARLY FINISHED 95

The rest is up to you 95

REFERENCES 97

Introduction

According to the Intergovernmental Panel on Climate Change (IPCC) report on the impacts of Global Warming published in April 2022, average global greenhouse gas emissions reached the highest levels in human history from 2010 to 2019, but there is increasing evidence of climate action and the rate of growth has slowed. Since 2010 there have been sustained decreases of up to 85% in the costs of solar, wind energy and batteries. An increasing range of policies and laws have enhanced energy efficiency, reduced rates of deforestation and accelerated the deployment of renewable energy,

The world has already warmed by 1° Centigrade since the 1880s, could reach 1.5°C by 2050 and is on track to reach 3°C unless we take action to reduce pollution and the emission of greenhouse gases. The impact of just a 0.5°C increase is very high, including more heatwaves, less freshwater, more intense storms, agricultural and food supply problems, sea level rises and the loss of virtually all coral reefs.

In the scenarios assessed by the IPCC, limiting warming to around 1.5°C (2.7°F) requires greenhouse gas emissions to peak before 2025 at the latest and be reduced by 43% by 2030; at the same time, methane usage would also need to be reduced by about a third. Even if we do this, it is likely that we will temporarily exceed this temperature threshold but could return below it.

The global temperature will stabilize when carbon dioxide emissions reach net zero. For 1.5°C, this means achieving net zero globally in the early 2050's: for 2°C (3.6°F), it is the early 2070s.

The headline of the report summary on the IPCC website is this; 'The evidence is clear: the time for action is now. We can halve emissions by 2030.'

Following on from this was the United Nations Climate Change Conference COP27 held in Sharm El Sheikh during November 2022. Although little progress was made in getting countries to make further commitments in the efforts to reduce global emissions, issues of critical importance to developing economies and those most at threat from climate change were at the forefront and genuine progress has been made.

In December 2019 Prince William announced the launch of the Earthshot Prize after two years of development inspired by his father (now King Charles), Grandfather Prince Philip and the broadcaster Sir David Attenborough. For ten years they will be awarding a grant of £1 million to the winners in each of five categories to enable them to continue and develop their environmental work. Based around the UN Sustainable Development Goals, the five categories are 'the restoration and protection of nature', 'air cleanliness' 'ocean revival', waste-free living' and 'climate action.' Aimed at bringing hope and positivity to what can be a depressing and worrying subject for many people, and inspiring the younger generations to get involved, the first prizes were awarded in 2021.

A record amount of renewable energy capacity was installed in 2021, despite post-pandemic delays and a rise in the cost of raw materials. This trend has continued throughout 2022 and, as nations that that have relied on fossil fuels from Russia push ahead with increases to their renewable energy capacity in response to the war in Ukraine, 2023 and beyond should see significant progress. There has also been increasing digitalisation, heightened attention to electrification of transport, a number of jurisdictions pledging to become coal-free, new policies and partnerships on carbon pricing, and new initiatives and goals set by governments at all levels.

Several renewable energy technologies such as hydropower, bioenergy and geothermal power and heat have long been established as mainstream and cost competitive sources of energy. Solar power, wind power and hydrogen are joining them: they are now competitive with new fossil fuel capacity in an increasing number of locations, and are coming closer to being competitive with existing fossil fuel and nuclear power generation.

Growth in renewable energy deployment and output continues, particularly in the power sector, thanks to several factors including increasing access to finance, concerns about energy security, the environment and human health, growing demand for energy in developing and emerging economies, the need for access to electricity and clean cooking facilities, dedicated policy initiatives and ambitious targets.

The responsibility is huge and the timing is crucial. We all know (even those who don't want to admit it) that our current path of capital and carbon driven development is wrecking our environment, draining resources and endangering the other life that we share the planet with and rely on for sustenance.

We need to change what is considered worth paying for and invest in green and renewable technologies, even if that means moving away from current methodologies and economic practices that are increasingly based around centralising everything. Embracing and adapting to technologies based around empowering individuals and small communities, especially personal and small scale power generation, is essential. It is time to stop pitting the environment against the economy and make an economy based on the best ways to improve and maintain the environment. Instead of debating endlessly over long term predictions of what may or may not happen and how bad it might be, we need to concentrate on enacting solutions to immediate and evident threats, the most important being to stop polluting the environment.

In order to achieve the above, more people need to understand the technologies and options available and how to contact like-minded people, companies and institutions so that they can become involved and encourage others to do the same. That is what this book aims to do.

In researching, planning and writing this book I visited hundreds of websites and read all sorts of articles and publications. At the end of each chapter I have provided the names and links for the best sites and articles. This enabled me to concentrate on providing clear explanations of the history, techniques and research involved whilst adding the ability for the reader to access more in-depth analysis and be proactive in becoming involved with and supporting the transition from our current carbon based economy to a renewable based environmentally friendly economy.

To start with we need to understand why we need to generate so much power. Our civilisation has developed and grown dramatically over the last 200 years. The two main things that have driven this is the increasing use of electricity, and internal combustion engines, a lot of which are used to generate electricity.

Recommended further reading

IPCC — Intergovernmental Panel on Climate Change   www.ipcc.ch

The World Economic Forum (weforum.org)   www.weforum.org

The Earthshot Prize: Urgent Optimism & Action to Repair the Planet
    www.earthshotprize.org

ClimateChange | United Nations
    www.un.org/climatechange

Electricity

What is electricity?

To answer this question you need to understand the basic structure of atoms.

The nucleus of an atom consists of one or more protons (positively charged particles) and neutrons (neutral particles). Rotating in a stable orbit around the nucleus are electrons (negatively charged particles). The atomic number of an element is determined by the number of protons it contains and the atomic mass is the number of protons and neutrons.

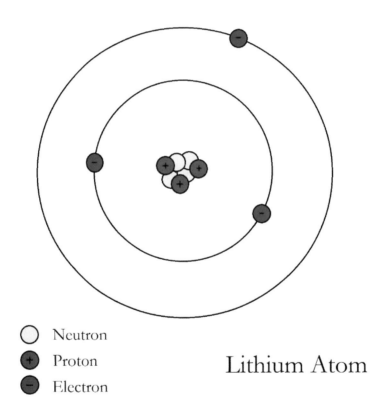

○ Neutron
⊕ Proton
⊖ Electron

Lithium Atom

Since opposite charges attract, protons and electrons are attracted to each other. Conversely, particles that have the same charge repel each other. Understanding these relationships quickly becomes extremely complicated, but luckily for us all you need to know at the moment is that the attraction and repulsion is what drives the movement of the particles, and the strength of that interaction is what makes atoms attract or repel each other.

The result of all this attracting and repelling interaction is the generation of forces.

A force is any pull or push that causes an object to alter its physical state. There are currently four recognised 'Fundamental Forces' in nature, although the latest research seems to suggest that there may be a fifth. The four Fundamental Forces confirmed so far are:

Gravitational Force

This is the weakest of the forces and acts on all things that have mass. The larger it is, the more force it exerts (so a large object attracts a smaller object). This is the force that holds us to the Earth and keeps the planets in orbit around the Sun.

Electromagnetic Force

This force acts on all things that have an electrical charge. Electrical energy comes from the movement of charged particles. Electric currents give rise to magnetism and magnetism can be used to induce electric currents.

The Weak Nuclear Force

This is responsible for particle decay and forms a part of nuclear fusion reactions, which is what powers the stars.

The Strong Nuclear Force

This is what binds the fundamental particles of matter together. Without this force, atoms would not be able to combine to form molecules.

The fundamental forces produce energy, which is defined as 'the ability to do work'. There are two general classifications for energy:

- *Potential Energy* is energy that is stored and has the potential to be released.
- *Kinetic Energy* is the energy produced by an object due to its motion.

The only fundamental force we are interested in here is Electromagnetism.

The presence of an electric charge, which can be either positive or negative, produces an electric field. The movement of electric charges is called an electric current and produces a magnetic field.

So, electricity is the energy generated by the flow of charged particles. It can be stored and used by manipulating the forces acting upon it.

Electricity in Human History

Pretty much everything relating to our current understanding of the Universe and how nature works started coming together during The Age of Reason in the 17th Century and continued to be defined, refined and used throughout The Renaissance and The Age of Enlightenment. The process continues apace in our modern world.

Electricity, being the most easily observable of the fundamental forces, has been known about since the dawn of civilisation, most obviously in the form of lightning.

The use of electricity to treat ailments and diseases is among the earliest recorded medical procedures. Dating back over 5,000 years, there are records of ancient Egyptian physicians using electric catfish from the Nile River to stimulate nerves in order to reduce the pain of arthritis.

The word 'electricity' is derived from the ancient Greek word 'elektron', meaning Amber (a fossilised tree resin. This is because they realised that by rubbing amber on fur you generate static electricity, causing attraction and sparks. The Greeks also used electric catfish for relieving headaches and gout.

In 1936, an artefact that came to be known as the Baghdad Battery was found in Iraq. Estimated to be about 2,250 years old, it is believed that it could have been used for electroplating, amongst other things.

There is also speculation that the light in the Alexandria Lighthouse, the ever burning lamp in the Temple of Isis and the Dundra light bulbs depicted in the Temple of Hathor in Egypt all point to ancient civilisations using electricity for lighting at the very least, but none of these hypotheses can be unequivocally proven.

Up until the mid-1700's scientists had known about and experimented with magnetism. This led to the development of compasses and began the evolution of ever more reliable navigation techniques, especially when combined with increasingly accurate timepieces.

Static electricity was also being studied, but without much success and very few useful observations until William Gilbert published De Magnete in 1600. This is a study of electricity and magnetism that distinguished the difference between magnetism and static electricity. He coined the new Latin word 'electricus', which gave rise to the English words electric and electromagnetism. Sir Thomas Brown used them to describe his own investigations into Gilbert's work.

Further studies were conducted by Stephen Gray, who is credited as the first researcher to make the distinction between conduction and insulation. He also discovered the phenomena of electrostatic induction during the early 1700's.

Charles Francois de Cisternay de Fay discovered the existence of positive and negative charges and their attractive and repulsive properties. He also expanded on Gray's work and published a scientific paper on the subject in 1734.

Electrostatic generators and Leyden jars

An electrostatic generator produces static electricity. These were gradually developed during the 1600's. Originally known as Friction Machines, they consisted of a glass globe or sphere which was rotated and rubbed against a woollen cloth. Initially these were used to make people's hair stand on end or to give them an electric shock.

A Leyden jar is a device that stores an electric charge. Originally they were glass jars filled with water. A wire suspended in the water ran through a stopper in the neck of the bottle and could be connected to an electrostatic generator.

The discovery of the Leyden jar is accredited to a Dutch scientist named Pieter van Musschenbroek, who was a Physics professor at Leyden (Leiden) University. He was following up on experiments performed by Ewald Georg van Kleist and Georg Matthias Bose in the mid 1740's.

The experiment was set up with a chain running from an electrostatic generator to a suspended metal rod. Holding a water filled bottle up to the rod in one hand, he received a powerful electric shock when he touched the rod with his other hand (thus completing the circuit).

Leyden Jars are the earliest form of electrical devices called Capacitors or Condensers, which store electrical energy in an electric field. They are used for stabilising voltage and power flow and are vital components in all electric and electronic devices.

One of the most famous people to experiment with Leyden jars connected in series was Benjamin Franklin. He is credited with coining the term 'Electric Battery' to describe a series of capacitors. He also showed that lightning is a form of electricity during a famous experiment when he flew a kite during a thunderstorm, which also led to the invention of lightning rods.

In 1780, Luigi Galvani and his wife discovered that the muscles in a dead frog leg twitched when struck by an electric spark. They are recognised as pioneers in the study of Bioelectricity.

Allessandro Volta was one of the first people to follow up on the Galvani's discovery and they disagreed over what was actually happening. He realised that the frog legs were conducting electricity caused by an electric current.

In 1794 he demonstrated that when two metals and a brine soaked cloth (or piece of cardboard) are arranged in a circuit they produce an electric current. In 1800 he used a stack of alternating metal discs separated by brine soaked cardboard to produce a current that flowed through a wire connected to the top and bottom contacts. This arrangement is now known as a Voltaic Pile, and was the first electrical battery capable of providing a continuous electric current to a circuit. It was also one of the first examples of electrochemical cells. Volta experimented with different metals and quickly settled on zinc and copper. A unit consists of three discs and by adding more units you increase the amount of electricity produced.

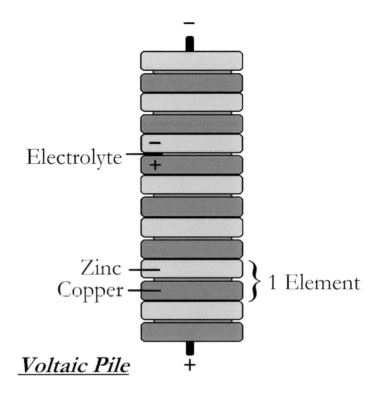

Voltaic Pile

Voltage, which is the measurement of the electrical pressure from a power source, is named after Volta.

Voltaic piles were very quickly used to make all kinds of scientific discoveries and world changing advancements, including telegraphy, electrolysis, photography, electric lights, and the linking of electricity, magnetism and light.

Sir Humphry Davy invented the first effective arc lamp in 1808. It used a piece of carbon that glowed when wired up to a battery.

In 1820, Hans Christian Orsted, A M Ampere and D F G Arago all confirmed the relationship between electricity and magnetism in separate independent experiments.

The electric motor was invented by Michael Faraday in 1821.

Then, in 1826, George Ohm defined the relationship between power, voltage, current and resistance. In 1827 he published his complete theory of electricity, which includes his law for electromotive force. This was the start of the subject of circuit theory. Ohm's law explains the relationship between voltage and the current flowing through resistors. The unit of electrical resistance is named after him.

The first dynamo (an electric generator capable of delivering enough electric power for industrial purposes) was built by Hippolyte Pixii in 1832. The French engineer and instrument maker constructed a hand cranked machine that rotated a magnet past a bar of iron wrapped with wire. This is one of the earliest examples of an alternating current, which he then converted to direct current.

In 1835 Joseph Henry invented the electrical relay, which is used to send electrical currents over long distances.

Thomas Davenport invented the modern electric motor in 1837. With help from his wife and brother he built his own batteries, electromagnets, and then a motor that ran on direct current. With support from professors and colleges in Vermont and upstate New York he eventually managed to obtain a patent from the U.S Patent Office, who had never given a patent on an electrical device before. He installed his motor in a small train operating on his property and set up the first printing press to run on an electric motor. He also invented the first electric streetcar and wanted his electric motors to replace steam powered locomotives, but the battery technology was not far enough advanced and he was never able to find any investors.

The fuel cell (a device that produces electrical energy by combining hydrogen and oxygen) was developed in 1839 by Sir William Robert Grove (this is covered in more detail in the chapter about hydrogen).

James Prescott Joule showed that energy is conserved in electrical circuits in 1841. This creates thermal heating and chemical changes. A unit of energy – the Joule – is named after him.

In 1844 Samuel Morse invented the Electric Telegraph, the first device capable of sending messages over long distances through wires.

During the 1860's James Clark Maxwell created a new era of physics. Maxwell's Equations (as they are now known) cover the laws of electrodynamics. They unify magnetism, electricity and light, and quickly led to the development and deployment of electric power usage and complex electronic devices including radios, televisions and computers.

In 1876 Charles Brush invented the 'open coil' dynamo, a generator that produces a steady current of electricity. Alexander Graham Bell (a Scottish inventor residing in the U.S.A.) invented the telephone in the same year.

Two years later the Englishman Joseph Swan invented the first incandescent light bulb (electric lamp), but it only had a short lifespan. At the same time, Thomas Edison founded the Electric Light Company in New York and began buying patents and developing practical long-lasting electric light bulbs.

Charles Brush also developed an arc lamp powered by a generator in 1878. The following year they were first used for public street lighting in Cleveland, Ohio. The California Electric Light Company Incorporated used two Brush generators to power 21 Brush arc lights in San Francisco and became the first company to sell electricity.

The first telephone exchanges were built in Britain in 1879. Then, on 28th December 1881, The Savoy Theatre in London became the first public building in the world to be lit entirely by electricity and received rave reviews the following day. Powered by an 88 KW generator, it used around 1,200 of Swan's Incandescent Lamps.

Swan was already using electric lights in his own property. In 1880 he had installed electric lights in his friend William Armstrong's house, Crayside. It was powered using hydro-electricity from a nearby stream.

Swan and Edison had registered very similar patents in 1879, and in 1883 the two competing companies merged to form The Edison and Swan United Electric Light Company.

On 12th May 1881, Werner Van Siemens completed the first official test of his 'Electric Streetcar', which was the forerunner of trams and electric railway locomotives. Running on a 2.5km line between Lichterfeide station in Germany and a nearby military academy, it was an instant success, transporting over twelve thousand passengers in the first three months of operation.

Nikola Tesla invented the Tesla Coil in 1883. A Tesla Coil is a transformer that changes electricity from low to high voltage. This was the initial step on the path to electricity being supplied over long distances.

In 1884 Tesla invented the Electric Alternator, which produced Alternating Current (AC) rather than the Direct Current (DC) produced by batteries. Generating AC electricity was another important advance on the path to enabling electricity to be supplied over long distances and the formation of electricity grid networks.

The first properly viable steam turbine was demonstrated in Newcastle, England by Charles Algernon Parsons in 1884. This made cheap and plentiful electricity available and revolutionised marine transport, naval warfare, and large scale electricity generation.

All of the basic knowledge, technology, techniques, resources and investment to provide, store and use electricity were now in place. Research, invention, development, refinement and deployment continued at a rapid pace and within fifty years peoples' lives were being transformed globally as the increasing use of electricity allowed unbelievably rapid progress in sanitation, medical care, hygiene, food preparation and storage, education, entertainment and commerce that continues today and shows little sign of slowing down.

Using Electricity

Actually using electricity is completely clean with virtually no pollution generated.

There are two main ways of using electricity.

The first is to take advantage of the effect it has on metals and some other elements. The most obvious use of this is in lighting and heat generation, but the most common use is in electronic circuitry.

The second method is to convert electrical energy into mechanical energy using electric motors.

Electric motors are an essential part of our everyday lives. Anything that spins, rotates or turns has an electric motor inside. The move to electric powered transport would not be possible without them.

All electric vehicles work in the same way – a source of electricity powers electric motors which turn wheels, fans and propellers.

The two main sources of electricity for transport are batteries and fuel cells, although trains and trams have been using electricity directly through power cables for over 100 years.

Generating enough electricity to power our modern society is where pollution really becomes a problem. The sooner we can move completely away from fossil fuel based power generation, the more chance we have of cleaning up the environment and reducing our impact on global warming and climate change.

How Electricity is produced, transmitted and stored.

Electricity is produced in two ways. Photovoltaic cells (Solar Panels), batteries and fuel cells generate electricity using chemical reactions. These are investigated in more detail later in the book.

By far the most common way of generating industrial scale amounts of electricity is by using steam, hot gases, moving water or the wind to spin a turbine.

Turbines

A turbine is a device that harnesses the Kinetic Energy of fluid movement and turns it into rotational motion.

Turbines use a set of blades that spin when a fluid passes over them, capturing some of the energy as rotational motion. The vast majority of electricity is currently produced by turning mechanical energy from a turbine into electrical energy using a generator.

In gas turbines, compressed air is heated and mixed with fuel. This mixture is then ignited, causing it to expand rapidly. It is then forced into the turbine, causing it to spin. This is the core of a jet engine, but gas turbines are also used to generate electricity. They use natural gas as the combustion fuel.

Hydroelectricity, tidal, nuclear, geothermal and concentrated solar power all use turbines to generate electricity and are covered in more detail later on.

Electricity transmission and distribution

The transmission and distribution of electricity are the two different stages involved in transferring electricity from a generating site or power plant to end users, typically our homes and businesses.

Transmission refers to the process of moving electricity in bulk from the generating site to electrical substations over long distances via heavy duty electrical cables. These transmission lines move large amounts of power at voltage levels that are much too powerful to be delivered directly to the end user. The high voltage has the effect of reducing the current, and in so doing lowering energy losses from resistance and heat generation. Most transmission lines use high-voltage alternating current (AC) but high-voltage direct current (HVDC) is also used as it can achieve better efficiency over large distances (hundreds of miles) and in underwater power cables. HVDC links are also used to help stabilise large distribution networks.

Distribution refers to the transfer of electricity from electric substations at levels that can be delivered directly to the end user. At a substation a step down transformer is used to drop the high voltage from the transmission line to primary distribution levels. Primary distribution is suitable for large consumers such as factories and industrial uses.

Smaller secondary distribution substations step down the primary levels further. Large properties, commercial and office buildings and small factories can take a supply voltage of 400 volts and residential properties are supplied at either 230 or 120 volts depending on which country they are in. (the U.K. uses 230 volts and in the U.S.A. it is 120 volts, for example).

Power losses during transmission and distribution

One of the biggest problems with distributing electricity through grid networks is the inherent power loss involved. A lot of the loss is caused by heat generation and general resistance. Generating electricity using steam powered turbines, no matter how the heat is generated, loses around 30 to 35% of the original energy potential. Once the electricity has been generated there are further losses in the transmission and distribution network.

1-2% of energy is lost during the step-up transformer process from when the electricity is generated to when it is transmitted.

2-4% is then lost in the transmission lines.

A further 1-2% is lost during the step-down process to transfer the electricity into the distribution system, which then loses another 4-6%.

This means that the overall energy loss between the power plant and the end user is usually in the region of 8-15%. One of the potential advantages of some renewable energy generating systems, including solar power and fuel cells, is that they can be used to generate onsite individual electricity and so do away with the need to have a transmission and distribution network. At the moment they are just on the verge of becoming practical for this due to the continuing increase in their energy generating efficiency.

So, now we understand the basics of electricity production, transmission and distribution, the next item on the agenda is how to store it. For most people alive today, it is impossible to imagine life without mobile energy storage. More than 15 billion single use batteries are sold globally every year to power small devices such as remote controls and flashlights.

The biggest developments have been in rechargeable batteries, both in technology and production capacity. Digital cameras, cell-phones, tablets, drones and all sorts of technological and mechanical devices, including electric vehicles, are only possible because of the huge advances being made in battery technology.

Recommended further reading

Turbine - Energy Education   www.energyeducation.ca

Electric power transmission - Wikipedia

Batteries

According to the dictionary, a battery is defined as a container consisting of one or more cells in which chemical energy is converted into electricity and used as a power source. All well and good, but how does that work in the real world?

Every battery is basically a Galvanic Cell where redox reactions take place between two electrodes, which are the source of the chemical energy.

A *redox reaction* is a basic chemical process whereby a reduction reaction and an oxidation reaction takes place.

Reduction happens when an atom gains one or more electrons. When an atom gains an electron, its negative charge increases.

Oxidation happens when an atom loses one or more electrons, thus losing negative charge and making the atom more positive. It is called oxidation because the first redox reactions observed were ones involving oxygen.

Essentially then, a redox reaction is a chemical reaction where electrons are transferred from one chemical to another. Rust is probably the most well-known oxidation reaction, and bleaching is an oxidation-reduction reaction.

Batteries can be broadly classified into two types – Primary and Secondary.

Primary Batteries

In primary cell batteries the redox reactions take place in one direction only. This means that they will only provide electricity until the chemicals inside it can no longer react. The batteries used in remote controls, torches and other small devices are examples of primary batteries. The most common chemical types are silver-oxide, alkaline, and carbon-zinc.

Secondary Batteries

Secondary cell batteries can be recharged by passing a current through the electrodes in the opposite direction (negative to positive). Rechargeable household batteries include cadmium and lithium-ion. The regular 12 Volt batteries in conventional internal combustion engine vehicles are lead-acid.

Major developments in battery technology

As outlined in the history of electricity, the groundwork that led to the development of modern batteries was made by Luigi Galvani and his invention of the Galvanic cell, followed by Alessandro Volta. His voltaic pile used zinc and copper for electrodes with brine soaked paper as the electrolyte. It was the first electrical battery that could provide a continuous electrical current to a circuit. He also disproved the then commonly accepted theory that electricity could only be produced by living organisms.

The main limit to the usefulness of voltaic piles (referred to as the "hydrogen bubble) was that bubbles formed on the bottom of the zinc electrodes thus interfering with and ultimately preventing the redox reaction.

The Daniel cell, first presented in 1836 by the British chemist John Frederick Daniel, solved the hydrogen bubble problem by using a copper pot filled with a copper sulphate solution. The electrical potential of the Daniel cell became the base measurement for voltage, rated at one volt.

In 1801 the physicist Johann Wilhelm Ritter further developed the Galvanic cell by working out how to make them rechargeable. Other people continued his work and in 1859 Gaston Plante constructed a lead-acid battery, recognised now as the first practical rechargeable storage battery. The basic construction and cell arrangement is still used today as they are ideal for turning starter motors for internal combustion engines and powering the vehicle's electrical systems.

Alkaline Batteries

Working independently of each other, Waldmar Jungner and Thomas Edison developed nickel-iron batteries that were first demonstrated in 1900. These batteries used alkaline electrolytes instead of acid based ones. Alkaline batteries proved to be more durable and could be used continuously.

The next significant advance was made by the Canadian engineer Lewis Urry in the 1950's, when he constructed the first zinc-manganese dioxide battery. The alkaline electrolyte used is potassium hydroxide. These batteries had increased shelf life and durability, which is the primary reason that they became the mainstay for all consumer electronic devices and enabled the development of the first transistor radios; followed by all consumer electronics. Duracell and Energizer are probably the most recognisable manufacturers of this type of battery.

Nickel-Cadmium (NiCad) Batteries

Ni-Cd or NiCad batteries are rechargeable and use nickel oxide hydroxide and cadmium as electrodes. The first NiCad wet cell battery was created by Waldmar Junger in 1899. Thomas Edison patented Nickel-Cadmium batteries in the U.S.A. in 1902. Development continued, but it was not until the 1950's that advances in manufacturing technologies made batteries cheaper to produce in large numbers.

Cadmium is a toxic heavy metal which is an environmental hazard and can be deadly to higher life-forms. Great care has to be taken in their manufacturing and disposal and they are gradually being phased out in favour of nickel-metal hydride (NiMH) batteries.

Nickel-Metal Hydride (NiMH) Batteries

Invented in 1967, NiMH batteries use hydrogen-absorbing alloys instead of cadmium. Their development was sponsored by Daimler-Benz and Volkswagen.

They became commercially available in 1989 and typically have two to three times the capacity of NiCad batteries. They also offer high-energy density, longer life, high power, durability and faster charging rates. They are now being replaced by the even more compact and durable Lithium-Ion battery.

Lithium-Ion (Li-Ion) Batteries

Following decades of research and development by various scientists and facilities, Sony released the first commercial Li-ion batteries in 1991. They have a much higher energy density than other batteries (meaning they can be lighter and smaller) and enabled the development of laptops, tablets, mobile phones and pretty much all modern consumer electronics and DIY tools. Battery electric vehicles only became viable because of them and they have the potential to be used as backup storage for national power grids. Slight variations in the composition of the anode and diode give different characteristics which can be adapted for various applications.

Li-ion batteries contain a flammable electrolyte and initially there were issues with them catching fire, but improvements to their design and added safety features have reduced this risk significantly.

The latest developments have enabled them to achieve up to 150 Watt-Hours (WH) of energy per kilogram, compared to NiMH at 70 WH/kg and lead-acid at 25WH/kg, They also have a lower discharge rate, typically losing about 5% per month compared to NiMH batteries, which lose around 20% per month.

Huge amounts of research and development is being done to improve the efficiency and storage capabilities of batteries. One interesting development is in saline technology, which uses salts as the electrolyte. At the moment they are expensive and do not have the energy density to challenge Li-ion batteries, but they are being assessed for large scale storage to support grid systems and you will be hearing more about them in the future.

The impact of batteries on the environment

Battery production uses a lot of energy, from mining the raw materials to the electricity used in the manufacturing process.

The negative environmental effects of mining include land, air, soil, and water pollution. It produces hazardous and toxic waste and can leach toxic chemicals, heavy metals and pollutants into the environment. This can lead to air pollution, the release of greenhouse gases such as methane and CO_2, and the leaching of toxic chemicals into freshwater sources such as lakes, streams, rivers and oceans.

Mining also uses a lot of water and electricity and can cause deforestation and loss of biodiversity. Mine workers' lives, health and safety can be at risk from exposure to hazardous air and chemicals. Mining has improved over the last few decades in developed countries, but some developing countries lag far behind in making improvements.

Batteries thrown away as household rubbish end up in landfills. As the casing corrodes, chemicals leach into the soil and pollute the environment as they make their way into the water tables, rivers, and eventually the oceans. Lithium can cause landfill fires that burn underground for years and release toxic chemicals into the air.

Whilst rechargeable batteries are considered to be the best option for long term use, single use alkaline batteries no longer contain large amounts of mercury and are considered safe to throw away in most countries, although you should always look for local recycling facilities and other safe disposal options.

Producing a battery electric vehicle can contribute on average twice as much to global warming potential and uses twice as much energy as producing an internal combustion engine vehicle, mainly due to the battery production. Once in use, its carbon footprint depends on how the electricity used to recharge the battery is generated. It is estimated that on average it will take around five years or 75,000 km of use to become less polluting than a vehicle powered by a fossil fuel internal combustion engine. Of course these figures vary depending on who is funding the research and how accurate the original data is. A more positive fact for consideration is that the pollution is moved away from city environments, which means improvements in air quality and reductions to health problems, especially those related to breathing.

So, in order to reduce the emission of greenhouse gases and pollution of the environment, we need to find cleaner and more efficient ways of generating and distributing electricity and powering our vehicles. This is what renewable energy is all about.

Recommended further reading

The Evolution of Battery Technology (visualcapitalist.com)

LION Smart - The History of Battery Technology: Evolution of Energy Storage

Electric vehicles guide - EV charging, models and costs UK 2023 (zap-map.com)

H is for Hydrogen

What is hydrogen?

The term 'Hydrogen Economy', relating to using hydrogen as the primary fuel for vehicles and power generation was reportedly first used by John Bockris during a talk given to the General Motors Technical Centre in 1970 (although the concept had been proposed back in 1923 by the British-Indian scientist J.B.S. Haldane). Bockris went on to publish the book 'Energy, the Solar Hydrogen Alternative' in 1975.

In this section you will find out why hydrogen has the potential to replace fossil fuels, how important it has been throughout history, and all of the various techniques involved in producing, storing and using it.

To start with we have to understand some basic concepts. Firstly we need to add a bit more detail to our understanding of atoms.

Atoms are the smallest parts of ordinary matter that have the properties of a chemical element: basically everything we can see, hear, touch or smell is made of atoms.

Atoms consist of a nucleus and one or more electrons (negatively charged particles) that rotate around the nucleus in various stable orbits. The nucleus is made of one or more protons (positively charged particles) and neutrons (neutral particles). The only exception is hydrogen, which only has one proton, all by itself. Hydrogen, therefore, is the simplest kind of atom there is. The chemical symbol for hydrogen is 'H'.

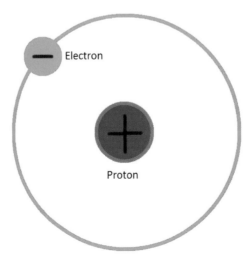

A hydrogen atom

According to the latest scientific research, hydrogen was the first atom to form following the Big Bang that created the universe. In cosmology this is called the Recombination Epoch, during which charged electrons first became bound to protons to form electrically neutral hydrogen atoms. According to the theory, this happened around 378,000 years after the Big Bang.

Hydrogen is now the most abundant element in the universe, making up about 75% of normal matter by mass and 90% by number of atoms. It is the fusion of hydrogen nuclei by thermonuclear or fusion reactions that is the primary source of energy generation in stars. The scientific term for this is 'hydrogen burning', although nothing is burnt in the ordinary sense of the word. The hydrogen atoms are squashed together under increasingly immense pressures until they fuse with each-other and create the more complex atoms of the other elements, starting with helium. This releases huge amounts of energy, mostly as heat and light.

Depending on the mass and age of the star, the fusion process becomes more complex and produces all of the known elements. During the various phases of the star's life cycle, different amounts of each element are produced and released into the universe. The mass of the star partly defines the way it will develop. Some don't have enough mass to explode: our own sun is one of these. Eventually it will consume all of the hydrogen in its core. With the hydrogen fusion stopped, gravity compresses the star. This increases the temperature and pressure until it reaches the point where helium begins fusing to form carbon, which then releases enough energy to overcome the force of gravity and the star expands outwards to many time its original size, becoming a red giant.

Red giants are cooler because the matter and energy occupy a much greater volume. Helium fuses into carbon and oxygen in the core. If it has insufficient mass to generate the one billion degrees Kelvin required to fuse carbon, an inert mass of carbon and oxygen builds up in its centre. Gravity takes over again and the star shrinks until a new helium shell is formed at the core. When the fusion process starts up once more, the energy blows the outer layers of the star away to form huge clouds of dust and gas known as planetary nebulae. This leaves behind a core, which is known as a white dwarf.

White dwarf stars are thought to be the final stage in the life cycle of stars that do not have enough mass to become neutron stars. Around 97% of all the stars in our galaxy fall into this category, including our sun. White dwarf stars are very dense and consist mostly of carbon and oxygen. They do not generate fusion reactions and are very hot to start with, but gradually lose this by radiation over billions of years.

Neutron stars are the ultra-dense remnants of stars that had enough mass to explode in what is called a supernova. There are two types of supernova.

A *Type 1 supernova* happens when a star accumulates matter from another nearby star until a runaway nuclear reaction ignites, causing a massive explosion.

A *Type 2 supernova* follows the same process as other stars until the hydrogen and helium are depleted. Because of its larger mass it has enough pressure to fuse carbon. Gradually the heavier elements are formed and build up in layers, with the heaviest elements towards the centre. Once it reaches a certain mass the star begins to implode. This makes the core become hotter and denser until the pressure in the core is so high that the implosion bounces off and ejects the surrounding stellar material into space.

So, all of the elements are formed in stars, and the process begins with hydrogen. The next thing that happens is that the atoms are attracted to each other and are able to bond together to form chains, which are called molecules. These can be composed of one element, or an increasingly complex chain of combinations of elements. Hydrogen atoms are part of the molecules that make life as we know it possible. They combine with oxygen to make water (two hydrogen atoms bonded to one oxygen atom) and with carbon to create hydrocarbons, which is what all living things are made of.

The elements also have variations in their atomic structure called isotopes. These are atoms of an element that have the normal number of protons and electrons but a different number of neutrons. Isotopes have the same atomic number and chemical properties but with a different mass. The standard isotope of hydrogen is called protium. There are two other isotopes of hydrogen that you need to know about.

Deuterium (chemical symbol ^2H), also known as 'heavy hydrogen,' has a neutron as well as a proton. Deuterium is about twice as heavy as protium.

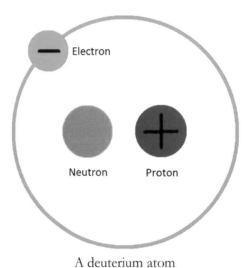

A deuterium atom

There are an average of 26 deuterium atoms for every one million hydrogen atoms. Water made from deuterium and oxygen is called heavy water (deuterium oxide, chemical symbol ^2H_2O) and has different nuclear, physical and chemical properties compared to H_2O. This makes it very useful as a reaction moderator and coolant in nuclear reactors because it absorbs less neutrons than normal water, which means that natural uranium can be used for the nuclear reaction instead of enriched uranium.

Deuterium is also used as a non-radioactive label in chemical experiments and has other uses in scientific experiments and studies, including techniques to produce nuclear fusion.

The other hydrogen variant is *tritium* (^{3}H). This atom contains two neutrons in addition to the proton, and is radioactive.

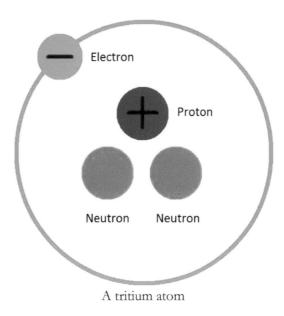

A tritium atom

Tritium is used to make luminous paint because the electrons released by radioactive decay react with phosphors to make them glow (replacing radium which has been banned because it can cause bone cancer). It is also used in self illuminating lights found in wristwatches, dials, compasses and sights for weapons. Other uses include nuclear fusion and fission experiments (with deuterium) and as a radioactive tracer in chemical and biological experiments.

Recommended further reading

Jefferson Labhttps://education.jlab.org/itselemental/ele001.htm

World of moleculeswww.worldofmolecules.com/elements/hydrogen.htm

Chemicool FactsHydrogen Facts – 12 of the best from Chemicool.com

Hydrogen in human history

As our modern civilisation gradually emerged from the Dark Ages and became more cohesive and structured, so did the development of science and the modern scientific methodology. The modern scientific method involves making observations, formulating hypotheses, testing and experimenting, gathering data, and developing theories. The results of all this have to be consistent and repeatable so that others can produce the same predicted results.

The modern scientific method began to really take shape and become universally accepted during the 17th century, when there was a lot of interest in and study of the Greek philosopher Aristotle's legacy from the 4th century BC.

As with all human development, there are lots of individual contributions along the way and then one person (or a specific group of people) suddenly puts all these things together to create something new, or build a solid platform that enables others to develop things further and faster.

Advances in technology and knowledge complement each other, and technical innovations bring about new discoveries that breed further discoveries and innovations whilst increasing our knowledge and understanding of nature.

The 17th century is often referred to as the 'Age of Reason' and is considered as the successor to the Renaissance and predecessor to the Age of Enlightenment. Work done in many fields, including philosophy, opened the path for more decisive steps towards modern science which took place during the 18th century, known as 'The Age of Enlightenment'. During the 19th century, the importance of science was recognised further and it became a proper profession with more institutions being established. This is a process that continues to this day.

17th Century – The Age of Reason

Towards the end of the 16th Century, improvements in the manufacture of optics and lenses enabled the invention of the microscope (1595), followed by the telescope in 1608. Francis Bacon postulated the basics of the modern scientific method. His work entitled 'Novum Organum Scientarium' (New Instrument of Science) was published in 1620.

In 1625, the Flemish chemist, physiologist and physician Jan Baptist van Helmont provided the first description of what we now call hydrogen. He is credited with introducing the word 'gas' (this is either from the Greek word for chaos - χάος (cháos), or the word 'ghast' (or 'gheist') meaning ghost or spirit – no-one can prove it either way). Van Helmont is also considered to be the founder of 'Pneumatic Chemistry', which began as a study of the physical properties of gases and how they relate to chemical reactions. This led to improvements in laboratory techniques and a greater understanding of the Earth's atmosphere. Later on it helped establish atomic theory, which in turn led to a greater understanding of atoms and molecules and how to measure their mass.

A key figure in the development of the scientific method was Rene Descartes, a French philosopher, mathematician and scientist. His highly influential works 'Discourse in Method' and 'La Geometrie' were published in 1637. In 1660 The Royal Society was founded in London and quickly became the British and Commonwealth's Academy of Science.

Largely regarded as the first modern chemist, Robert Boyle published what is now known as 'Boyle's Law' in 1662. This describes how the pressure of gas tends to increase as the volume decreases. He was a founder member of The Royal Society and was elected as a Fellow of The Royal Society (FRS) in 1663. In 1670 he produced hydrogen by reacting metals with acids and published 'New Experiments Touching the Relation between Flame and Air' in 1672.

18th Century – The Age of Enlightenment

Experimentation and discovery continued apace across all fields of science and in 1766 Henry Cavendish was the first to recognise hydrogen gas as a distinct or 'discrete' substance. He called it 'Flammable Air'. In 1781 Cavendish discovered that hydrogen gas produces water when burnt, and is usually given credit for its discovery as an element.

The influential French chemist Antoine Lavoisier gave the element the name Hydrogen (from the Greek words hydro (water) and genes (creator) when he and the French scholar Laplace reproduced Cavendish's findings in 1783.

Also in 1783, the French inventor and scientist Jacques Charles, having studied the work of Robert Boyle and his contemporaries, worked in conjunction with the Robert brothers to launch the first unmanned hydrogen balloon. The experiment was a resounding success and they quickly followed it up by launching the first manned flight, which lasted for over two hours and reached a height of about 550 meters (1,800 feet). More flights followed and ballooning quickly gained mass popularity. This led to increased research into ways of producing, manufacturing and storing hydrogen. It is also reflected in various articles and artwork speculating on the future, and early science fiction stories, including some of Jules Verne's writings.

19th Century

The next really important advance in science was the discovery and study of electricity. Alessandro Volta published his experiments in 1800. He wrote to the London Royal Society describing the technique for producing electric current using his device the Voltaic Pile, which was the first electrical battery that could produce a continuous electrical current to a circuit. In the same year it was used to discover how to perform electrolysis of water, in which an electric current passed through water splits it into its component gases of hydrogen and oxygen. Over the next decade the Cornish scientist and chemist Humphry Davy used the Voltaic Pile to discover and isolate the chemical elements sodium, potassium, calcium, boron, barium, strontium and magnesium.

Then, in 1806, Francois Isaac de Rivaz built the de Rivaz engine, generally acknowledged as the world's first internal combustion engine (ICE). De Rivaz had served in the French Army and his experience with cannons led him to think about using an explosive charge to drive a piston instead of using steam pressure. His first design was a stationary single cylinder engine used to drive a pump. It was fuelled with a mixture of hydrogen and oxygen which was ignited by an electric spark in the same manner as a modern ICE. In 1807 he placed his experimental engine in a carriage and made the first ICE driven vehicle. Unfortunately few of his counterparts took his work seriously (although there were others experimenting along the same lines). The French Academy of Sciences argued that the internal combustion engine would never rival the performance of the steam engine. (This kind of thing happened more often than you would think – most famously when Lord Kelvin (one of many sceptics) stated that 'heavier than air machines are impossible' and was proved wrong by the Wright Brothers eight years later.

The famous scientist and Science Fiction author Arthur C Clarke commented on this by stating, 'When a distinguished but elderly scientist says that something is possible, he is very likely right. When a distinguished but elderly scientist says that something is impossible, he is very likely wrong.'

In 1819, hydrogen was first used in a blowpipe by Edward Daniel Clarke, an English Clergyman, naturalist, chemist, mineralogist and traveller. Blowpipes are tools that channel a concentrated jet of gas through a flame to create a very hot jet. Contemporary blowtorches and welding and cutting torches are modern developments of the blowpipe. Clarke mixed the hydrogen with oxygen which vastly increased the temperature and drove the development of the gas blowpipe as a tool. This very quickly enabled its use in exploring the characteristics of heating a much larger range of materials to high temperatures.

Limelight was discovered using the oxy-hydrogen blowpipe to rapidly heat a cylinder containing quicklime. This creates an intense illumination that is most famous for its use as stage lighting. One of its earliest uses was for indoor stage illumination at the Covent Garden Theatre in London in 1837. It became widely used in theatres around the world until it was replaced by electric arc lights in the late 1800's.

The first references to fuel cells appeared in 1838 and 1839 in letters published in 'The London and Edinburgh Philosophical Magazine and Journal of Science'. Two letters came from Welsh physicist William Grave and a third was from the German physicist Christian Schonbein. They discussed electric current generated from hydrogen and oxygen dissolved in water. Grove later supplied a sketch of his fuel cell that used similar materials to today's phosphoric acid fuel cells

In 1866 August Wilhelm von Hoffman, a German chemist who became the first director of The Royal College of Chemistry in London in 1845, invented the Hoffman Voltammeter, an apparatus for electrolyzing water.

Hoffman voltameter

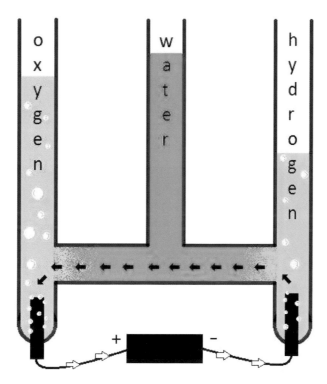

When an electric current is run through the voltammeter, gaseous oxygen is formed at the anode (positive terminal) and hydrogen is formed at the cathode (negative terminal). Each gas displaces water and collects at the top of its tube. By comparing the amount of each gas collected from a volume of water, the composition of water (H_2O) is proven as the gas volume is 2:1 (twice as much hydrogen as oxygen). This has become a classic demonstration experiment and I have seen it both in chemistry lessons and on TV programs.

In his 1874 novel 'The Mysterious Island', Jules Verne wrote 'Water will one day be employed as fuel, in that the hydrogen and oxygen of which it is constituted will be used.'

French chemist Paul Sabatier defined what is now known as the Sabatier Process in 1897. This process is the basis of hydrogenation (a process explained a bit later in 'Practical Application and Uses of Hydrogen'). Sabatier shared the 1912 Nobel Prize in Chemistry for his pioneering work in this field.

In 1898 the Scottish chemist and physicist James Dewar liquefied hydrogen using a regenerative cooling system and invented the Vacuum Flask. Unfortunately for him, he did not patent it and so was unable to prevent the company Thermos from using his designs, and did not profit from its widespread use. He next experimented with a high powered hydrogen jet and found that extremely low temperatures could be achieved by the resulting Joule-Thomson effect (very basically the heating or cooling of gas during expansion, but far too technical to explain further here). The success of these experiments enabled him to have a large regenerative cooling machine at the Royal Institution and in 1899 he collected solid hydrogen for the first time by reducing its temperature to -259.16 °C (14.01 °K).

Airships

The 20th century started with the launch of the first hydrogen filled airship by Count Ferdinand Von Zeppelin, a former general in the German army. On retiring from the army he followed up his interest in the emerging flight technology of balloons and founded the Zeppelin Airship company. The Zeppelin LZ1 was the first successful rigid airship. It did not have enough of an impact to attract investment from the government but, with the support of the King of Wurttemburg (whom he had served as an adjutant during his army career) and a contribution from Prussia, Zeppelin was able to continue developing airships.

The LZ2 made only one flight due to a series of mishaps and technical problems. In 1906 the LZ3 made two successful flights and in 1907 flew at a speed of 36 miles per hour. This achievement caught the attention of the Reichstag, and increased the growing public interest in airships. A larger model, the LZ4, was destroyed by fire after breaking free of its moorings during a storm. The LZ3 was revamped and, with the Kaiser's brother as a VIP passenger, flew non-stop for five hours and fifty-five minutes on the 27th of October 1908. On the 7th of November it flew a distance of 80 km (50 miles) and on the 10th of November Zeppelin gave a short demonstration flight to the Kaiser and was awarded the Order of The Black Eagle.

Zeppelins were built for the army and the company made a decision to capitalize on public enthusiasm by establishing a passenger carrying business. By 1914 the German Aviation Association had transported 37,250 people on over 1,600 flights without any incidents, and the Zeppelin revolution began creating the age of air transportation.

On the 2nd of July 1919, the British made airship R34, which was the length of two football fields. It left East Fortune in Scotland and completed the first transatlantic airship flight when it arrived in Mineola, New York, four and a half days later. The R34 was refuelled and turned around for the return flight, which left on the 10th of July and took seventy-five hours to fly to Norwich, England

The Zeppelin LZ129 Hindenberg was destroyed by fire in 1937. News of the tragedy rapidly spread around the world thanks to the now famous newsreel coverage and this rapidly brought about the demise of the airship as a commercial transport.

In 1997, retired NASA engineer Addison Bain challenged the belief that hydrogen caused the Hindenburg accident. The hydrogen, Bain demonstrated, did not cause the catastrophic fire. The real culprit was the combination of static electricity and highly flammable material on the skin of the airship.

20th Century

The German chemist Wilhelm Norman introduced the process for the hydrogenation of fats in 1901. This converts liquid vegetable oils into solid or semi-solid fats, such as those present in margarine. These fats are cheaper to produce than animal fats and were quickly picked up by food manufacturers, especially for baking, when it was discovered that they improve the shelf life of the product (this is explained in more detail under Practical Applications and Uses of Hydrogen).

In 1903 the British engineer Howard Lane invented the Lane Hydrogen Producer. The first commercial version was commissioned a year later and by 1913 twenty-four million square meters (850,000,000 cubic feet) of hydrogen was being produced annually, a proportion of which was used in airships. The process continued to be used until the 1920's when it was superseded by cheaper methods.

During the early part of the Twentieth Century there was a lot of research being carried out investigating the spectrums of gases, beginning with hydrogen. This research led to major advances in physics and the development of Quantum Mechanics.

Throughout the 1920's and 30's, new methods of producing hydrogen and using it for commercial purposes were developed. These methods include steam reforming and hydrocracking, which are explained in more detail later on in this book.

Having studied the explosive reaction of hydrogen and oxygen, the English physicist Cyril Norman Hinshelwood first described the chain reaction phenomena in 1926.

In 1930 the German engineer Rudolf Erren patented the Erren engine in the U.K. It was described as relating to the use of a mixture of hydrogen and oxygen as a fuel for internal combustion engines. The U.S. patent was registered in 1939.

The Heinkel HeS1 hydrogen fuelled centrifugal jet engine was tested in 1937 and is recognized as the first working jet engine. In the same year, the first hydrogen cooled turbo generator went into service in Dayton, Ohio, U.S.A. Today this method for cooling steam turbines is the most common type in its field.

The Rhine-Ruhr pipeline, completed in 1938, was the first hydrogen pipeline (used to transport hydrogen from the point of production to the point of demand). It runs for one-hundred and fifty miles (two hundred and forty kilometres) and is still in use today.

Liquid hydrogen was tested as a rocket fuel at Ohio State University in 1943.

Nuclear Weapons

The first nuclear weapon to be developed was a fission bomb, more commonly known as the atomic or A-bomb. The fission process uses heavy radioactive materials (usually uranium or plutonium isotopes) because the unstable atoms decay and emit neutrons that strike other atoms, causing them to split apart. This releases more neutrons and the result is a release of energy. Under controlled conditions, this method of releasing energy is what powers a nuclear reactor.

Fission bombs use conventional explosives to force the radioactive material into a tight, extremely dense ball until it reaches critical mass, which is when the fission process starts. The core heats up exponentially until it reaches supercritical mass. When this happens it triggers a chain reaction that instantaneously releases energy equivalent to tens of thousands of tons of TNT.

The first test detonation of a fission bomb was made by the Allies in Alamogordo, New Mexico on 16th July 1945. Codenamed Trinity, it marked the successful conclusion of the Manhattan Project, which was the Allies' Top Secret nuclear weapon development project. Three weeks later, on 6th August 1945, a uranium fission bomb was detonated over the Japanese city of Hiroshima, and three days after that a plutonium device was dropped on Nagasaki. The unprecedented destructive capacity of these two bombings and the Soviet Union's declaration of war with Japan on 8th August brought about the announcement of Japan's surrender to the Allies on 15th August. This was officially ratified when the Japanese government signed the instrument of surrender on 2nd September, effectively ending World War II.

Although they definitely speeded up the surrender of Japan and the cessation of hostilities, the justification for the atomic detonations in Hiroshima and Nagasaki is still heavily debated.

A second type of nuclear weapon was subsequently developed. Using fusion instead of fission, these are referred to as hydrogen or H-bombs. The process involves using a fission bomb as a trigger to heat a charge consisting of the isotopes of hydrogen to a temperature of millions of degrees. The hydrogen isotopes fuse into helium and that sets off a chain reaction resulting in an explosion many times larger than a fission bomb can achieve – in the order of megatons of TNT.

Ivy Mike, the code name for the first hydrogen bomb, was tested in 1952 but was quickly surpassed by a non-cryogenic (dry fuel) hydrogen bomb codenamed Shrimp. Tested in 1954, Shrimp produced a much more powerful explosion than anticipated and caused the most significant radioactive contamination accident in U.S.A. history.

Post War

During the late 1950's, Pratt and Whitney developed their model 304 jet engine using liquid hydrogen fuel as part of the Lockheed CL-400 Suntan project, intended to create a faster successor to the U2 spy plane with increased altitude capability. The project was unsuccessful due to the infrastructure costs involved, but the plane research led directly to the conventionally powered and highly successful SR71 Blackbird. The fuel research laid the groundwork for the use of liquid hydrogen combined with liquid oxygen as a rocket fuel. This enabled the development of ever more powerful rockets that were used for the Mercury, Gemini and Apollo projects and, in fact, nearly all commercial and scientific rocket launches since. It wasn't until 2007 that other rocket propellants became viable.

The RL-10 liquid hydrogen/oxygen fuelled rocket engine made its first successful flight in 1961, and the first commercial use of hydrogen fuel cells was in project Gemini, the spaceflight proving missions that tested all of the procedures needed for the Apollo moon landings.

In 1958 the specifications for a double-axle liquid hydrogen powered semi-trailer were issued and over the next few years the first hydrogen fuel cell tractor, fork lift truck and welding machine were built and demonstrated.

The American engineering company Allis-Chambers built a 750 watt fuel cell powered, one man underwater research vessel in 1964 and the first fuel cell powered golf carts a year later.

In 1966, U.S. car manufacturer General Motors introduced the Electrovan, which was the first time fuel cells were used to drive the wheels of an automobile (in this case a modified Handivan). Unfortunately it was not economical to mass produce as it used platinum as the catalyst in the fuel cells, and there were a host of other problems with the prototype.

Also in 1966, it was discovered that by reducing the temperature of hydrogen to around its melting point from frozen you can produce 'Slush Hydrogen'. This is 16 to 20% more dense than liquid hydrogen and was proposed as a rocket fuel to replace liquid hydrogen in order to improve the storage capacity by volume and reduce the launch weight of the vehicle. This means you use less space (volume) to carry the amount of fuel required to achieve orbit.

The liquid hydrogen/oxygen rocket engine J2 was first used in 1967. It powered the Saturn 1B and Saturn V rockets and was designed to be shut down and re-started. The first burn shut down after placing the vehicle in Earth orbit and then the rocket was reignited for a second burn to launch the vehicle into a lunar trajectory.

In the same year, the Japanese chemist Akina Fujusima was working on his Ph.D. under the supervision of Professor Kanichi Honda when he discovered the phenomenon of photocatalytic water decomposition (water photolysis) by exposing a titanium dioxide electrode to strong light. This is now called the Honda-Fujusima effect and is used for photo catalysis in photo electrochemical cells (PECs). These produce electrical energy (or hydrogen) in a process similar to the electrolysis of water. It has been referred to as artificial photosynthesis and could be a way of storing solar energy as hydrogen for use as a fuel. Akina Fujusima has had a long and distinguished career and was President of the Tokyo University of Science from 2010 to 2018.

The Hydride compressor was also demonstrated for the first time in 1967. It is a method of long term hydrogen storage for use in spacecraft and automobiles.

In 1971 the nickel-hydrogen battery was patented. This is a rechargeable battery and is used in the Hubble Space Telescope, the International Space Station, and on many space probes including the Mars Global Surveyor.

The European Space Agency's first launch system Ariane 1 made its debut flight on 24th December 1979. It used HM7B liquid hydrogen rocket engines to launch their CAT-1 technological test capsule into orbit.

NASA launched the first Space Shuttle in 1981. The Shuttle main engine was a liquid hydrogen and oxygen fuel cryogenic rocket engine. The Shuttle also used alkaline fuel cells to provide electricity, and the water produced was used for the crew to drink, and as a coolant in heat exchangers.

In 1997 Anastasios Melis, an American biologist at USC Berkeley discovered that depriving algae of sulphur makes it switch from producing oxygen to producing hydrogen. This is covered in more detail a bit later in the book.

German car manufacturer Daimler-Benz and Ballard Power Systems announced a $300-million research collaboration on hydrogen fuel cells for transportation.

21st Century

The first type IV hydrogen tanks for storing compressed hydrogen at 700 bar (1,000 psi) were demonstrated in 2001.

In 2002 the first 'Hydrail' locomotive was demonstrated in Val d'or, Quebec. It was a 3.6 tonne, 17kw hydrogen powered locomotive. A 380 passenger train was demonstrated by CSR Sifang co ltd in Qingdao, China in 2015.

Deep C, an autonomous underwater vehicle propelled by an electric motor powered by a hydrogen fuel cell was debuted in 2004.

The Ionic Liquid Piston Compressor was first demonstrated in 2005. Used to compress hydrogen up to 1,000 bar (1,450 psi) in hydrogen fuelling stations, it is much less complex than conventional piston compressors, has a much longer service life (as much as 10 times more) and is around 20% cheaper to run.

By 2007, fuel cells had been developed to the point where they complied with the codes and standards of their target markets and began to be sold commercially, initially as Auxiliary Power Units (APU) and backup generators.

Honda began leasing their FCX Clarity fuel cell electric vehicle in California in 2008.

By 2014 the Japanese hydrogen fuel cell micro combined heat and power (MCHP) ENE-FARM project (generating electricity for individual houses or buildings on site) had sold over 100,000 systems.

2015 saw the first commercial power to gas installation in Falkenhagen, Germany. This uses electricity to produce hydrogen, which is then stored in the natural gas grid.

London's first zero carbon hydrogen refuelling station was opened in Teddington at the beginning of May 2016. It features vehicles from Hyundai, Honda and Toyota and was the first of three stations to be opened under the HyFive project.

A company called Cemvita has been developing microbes that break down hydrocarbons and release hydrogen and carbon dioxide. They can be used in depleted oil wells and the first test programs have shown it to be a very efficient way of producing low cost hydrogen. The company are calling the end product 'Gold Hydrogen' and have already generated a lot of interest and investment.

Rolls Royce have successfully converted one of their jet engines to run on hydrogen. Easyjet and Airbus have partnered with them to develop new engines designed to run on hydrogen, and the aircraft to utilise them.

Recommended further reading

Hydrogen Power | The Water Page

How hydrogen is produced, processed and stored

Because hydrogen is the lightest element, any that gets released into the atmosphere quickly rises up and escapes into space. Therefore, to produce hydrogen you have find trapped hydrogen, release and capture it.

As hydrogen bonds so easily with most elements, there is a huge amount trapped in our environment, including just recently discovered underground reserves. It can be released from water using electrolysis and from hydrocarbons in many ways, the most common of which is currently steam reforming. It has also been confirmed that there are reserves of hydrogen gas underground. A town called Bourakebougou in Mali, West Africa is now powered by hydrogen from an underground reserve and the race is on to find more reserves, as this would be a viable, clean and extremely cheap way of producing as much hydrogen as we need.

So, hydrogen can be produced from diverse common resources including fossil fuels, biomass water, and underground reserves. The environmental impact of hydrogen production depends on the method used.

You will have heard a lot about Blue, Grey and Green hydrogen but for most people this has very little meaning. The colour refers to how environmentally friendly the process is. Below is a quick guide followed by a more detailed look into the methods used to produce hydrogen.

Green hydrogen

This refers to hydrogen produced by methods which produce no harmful pollutants or greenhouse gases. At the moment this means splitting water into hydrogen and oxygen using electrolysis.

The cost of producing green hydrogen is expected to decrease rapidly over the next decade. Just to confuse the issue there are subcategories designed to show where the electricity used for electrolysis comes from. Using wind power, hydro power and geothermal are classified as green. Other methods include Pink hydrogen (using nuclear power) and Yellow hydrogen (from solar power).

Blue hydrogen

Using steam reforming of natural gas produces blue hydrogen, but the by-product is carbon dioxide which needs to be captured and stored to avoid further greenhouse gas emissions. As green hydrogen production becomes more efficient and cheaper, blue hydrogen production will be phased out.

Grey hydrogen

This is currently the most common form of hydrogen production and is basically the same as blue hydrogen but without the carbon capture and storage.

<u>Black and Brown hydrogen</u>

Using black coal or brown coal (lignite) and gasification of fossil fuels is generally regarded as the most polluting method.

<u>White hydrogen</u>

Naturally occurring geological hydrogen (trapped below ground in geological rock formations mostly) can be released through methods such as fracking. This is likely to be renamed as Gold Hydrogen because it will probably be the cleanest method of producing hydrogen by a country mile. This is just starting to be properly explored. There are projects in Spain, South Korea, Nebraska U.S.A., and South Australia to investigate natural hydrogen resources, with many other areas under consideration. So far the signs are very encouraging and most of the exploration has shown that helium is also present in useful quantities.

Hydrogen production methods

<u>Steam reforming</u>

This is currently by far and away the most commonly used method of producing hydrogen and accounts for 85% to 90% globally. It is sometimes also referred to as Steam Methane Reforming (SMR).

High temperature steam and a methane source (most commonly from natural gas) are combined with a catalyst (usually nickel) at high pressure. This produces hydrogen, carbon monoxide and small amounts of carbon dioxide. The carbon monoxide is then reacted with steam and a different catalyst (copper or iron) at a lower temperature, in what is called a gas-shift reaction. This produces more hydrogen and carbon dioxide. The final process is called pressure-swing absorption and is used to remove impurities from the gas stream, leaving essentially pure hydrogen as the end product.

Steam reforming can also be used to produce other fuels such as methanol, propane and petrol. A major disadvantage of using this method is the amount of carbon dioxide that is produced. It is possible to use carbon capture techniques to reduce the amount of CO_2 released into the atmosphere but this increases the overall production cost and is not commonly used.

<u>Gasification</u>

Coal has developed a bad reputation over the last century because burning it produces a lot of particle pollution (the cause of chemical smog amongst other things) and greenhouse gases. It is, however, a complex substance that can be converted into a variety of useful products. Gasification is a process that produces power, liquid fuels, chemicals and hydrogen from coal and other sources including petroleum coke and a variety of biomass and waste-derived feed stocks.

This process is gaining in popularity for electricity generation but it has been used in the chemical, refining and fertiliser industries for decades because it produces a mixture of mainly hydrogen and carbon dioxide call Syngas, which is used to produce methanol, nitrogen based fertilisers and hydrogen. Even the glasslike by-product of the gasification process, called slag, can be used in roofing materials and road surfacing.

In a coal fired power plant, the coal is burned to produce heat which turns water into steam to drive a turbine. This releases a whole range of pollutants including large amounts of carbon dioxide. By comparison, the gasification process uses high pressures and temperatures which cause the coal to undergo different chemical reactions. It is sometimes referred to as 'clean coal' because it doesn't involve spewing toxins and CO2 into the atmosphere. The coal, however, still has to be mined and is not a renewable energy source. Biomass gasification produces the same results, but uses renewable materials such as wood, plants and organic waste as a fuel source.

Gasification was first developed by a Scottish engineer called William Murdoch in the late 1700's. He had spent most of his career developing steam engines before turning his attention to experimenting with the production and use of gases. His house in Redruth, Cornwall was the first domestic residence to be lit by gas. Apart from the benefits of gas lighting and heating, the process for producing Syngas also generated a number of other substances that were successfully exploited by other people. Among those substances were coke, ammonia, phenol (carbolic acid) and coal tar.

Phenol is a disinfectant and was used as one of the components in the first synthetic plastic – Bakelite. Coal tar contains a number of organic chemicals, including acetylsalicylic acid, more commonly known as Aspirin. Coal tar was also used to produce the first synthetic dye (mauve).

Cities in Europe and America began using Syngas (which quickly became known as Town Gas) to light their streets and houses. Eventually it was replaced by natural gas, and then electricity (generated from coal burning power generation (oh dear!)).

As concerns about pollution, climate change and power generation have increased over the last few decades, gasification has been making a comeback and is becoming more efficient all the time, to the point that it is now being touted as an important source of clean, renewable energy.

Electrolysis

This is the use of electricity to split water into hydrogen and oxygen. The reaction takes place in a unit called an Electrolyser, in a similar process to that used in fuel cells. They can be small units for localised hydrogen production up to large scale central production facilities. Depending on how the electricity is produced, this can be a zero emission production method.

Electrolyser

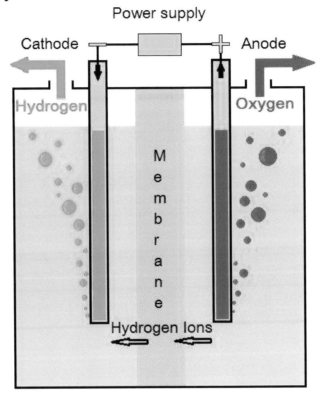

The main types of Electrolyser are as follows;

Polymer Electrolyte Membrane (PEM)

The electrolyte is a solid plastic based material. Oxygen is produced at the anode. The electrons pass through an external circuit and the hydrogen ions (photons) move through the PEM. At the cathode, the protons react with the electrons to produce hydrogen gas. PEM electrolysers typically operate at between 70°C and 90°C.

Alkaline Electrolyser (AEC)

Traditionally these use a liquid alkaline solution such as sodium or potassium hydroxide as the electrolyte. New technologies are now emerging that use solid alkaline membranes. As in PEMs, oxygen is produced at the anode and hydrogen at the cathode. They typically operate in the 100°C to 150°C temperature range.

Solid Oxide Electrolyser (SOEC)

These use a solid ceramic material as the electrolyte. In this method, water at the cathode combines with electrons from the external circuit to produce hydrogen gas and negatively charged oxygen ions. These pass through the ceramic membrane and react at the anode to produce oxygen gas and the electrons that flow through the external circuit to the cathode. To function properly, this process must operate at about 750°C. They can use heat produced by external sources to decrease the amount of electricity required to produce hydrogen from water.

Do try this at home

The basic method of electrolysis is extremely simple and you can do it at home. You need a nine volt battery, two paperclips and a container of water.

Unbend the paperclips (or use electrical wire). Connect one to the positive terminal of the battery and the other to the negative terminal. Then place the other ends in the container of water, as far apart as possible. You should see bubbles coming off of both wires in the water. The one with the most bubbles is producing pure hydrogen. The other is producing impure oxygen. You can test which is the hydrogen by lighting a match over the water (be careful). The hydrogen bubbles will burn whereas the impure oxygen will not. Using an electrolyte (such as salt) in the water increases the reaction but may produce small amounts of chlorine gas.

I tried this and the hydrogen gas formed quite quickly. The bubbles give a satisfying pop when lit with a match. There was very little going on at the oxygen end, but the wire oxidised quite quickly. When I used paperclips, the wire oxidised, turned brown, and dissolved when left overnight. Using copper electrical wire turned the water a bluish green colour as it oxidised.

Thermochemical cycle

Adding the prefix Thermo (from the Greek thermos, meaning heat) in front of the word chemical tells us that this is a process that uses heat to produce chemical reactions. The cycle part refers to the fact that the chemicals are reused in each cycle, creating a closed loop that only consumes water and produces hydrogen and oxygen.

The process requires high temperatures (between 500°C and 2,000°C) and is still being studied and developed (there are literally hundreds of different chemical reactions that can be used). It is hoped that providing the heat from, for instance, concentrated reflected solar energy will provide a clean method of producing hydrogen.

Photobiological

This is a way of producing hydrogen from green algae. It is a recently developed technology mostly resulting from research by Anastasios Melis, an American biologist at the University of California, Berkeley. He is currently Professor of Plant and Microbial Biology and Editor-in-Chief of the Planta Journal, a monthly peer-reviewed scientific journal covering all areas of botany.

In 1998, following research dating back to the 1930's, Professor Melis discovered that depriving particular strains of green algae of sulphur causes them to switch from producing oxygen to producing hydrogen. Work is continuing to improve the efficiency and this is a very promising way of producing hydrogen from sunlight with low to zero carbon emissions. The algae and bacteria could be grown in water that cannot be used for drinking or agriculture, and even waste water. Bioreactors are currently being developed to maximise the potential of the process.

Photocatalytic

This is a method of producing hydrogen by artificial photosynthesis. It uses a substance called a photocatalyst, and light. The latest research shows that solar water splitting using a powdered photocatalyst is an effective approach to producing hydrogen in a simple and inexpensive way.

There is a lot of research being done into alternatives to water, and improvements to the electrolysis process. Rice University's Laboratory for Nanophotonics, Syzygy Plasmonics Ltd and Princeton University's Andlinger Center for Energy and the Environment have been collaborating on research and development of "antenna-reactor" plasmonic photocatalysts which are showing huge potential. One of the most promising projects is the development of an Ammonia Cracking device that uses cheap materials, is small, lightweight, and runs at room temperature rather than the hundreds of degrees currently required. Ammonia is a carbon-neutral hydrogen compound that can be easily liquefied, stored and transported. It has a high energy density and hydrogen storage capacity, and there are many projects globally aimed at using it to produce green hydrogen.

Bacterial Biohydrogen

In a similar manner to that used in alcohol production, hydrogen can be produced by fermentation, which is a process that converts sugar to acids, gases, or alcohol. Using the recently developed and ever evolving device called a bioreactor, this advanced biofuel producing technology is becoming more attractive as biohydrogen is very easy to collect from the bioreactor.

To produce hydrogen, bacteria are used with organic acids that promote high production rates to speed up the process. The most important organic acids are acetic acid, butyric acid and propionic acid. These can all be derived from any organic material, including sewage waste water and agricultural waste.

So, what is so important about these acids? Well, let's take a closer look.

Acetic acid

This is produced and excreted by certain bacteria found universally in foodstuffs, water and soil. It is also produced naturally by fruit and other vegetation as it begins to decompose, and can be made synthetically using bacterial fermentation. You know it in a diluted form as vinegar, which is the oldest known application of acetic acid.

Acetic acid is also used in an extremely wide variety of applications and chemical processes. In particular, it is used to produce Vinyl Acetate Monomer (VAM), which accounts for about 45% of the global production of the acid.

VAM is used to make PVA, a type of thermoplastic used extensively in the manufacture of glues, adhesives and paints. It is also used to produce solvents, textiles, photographic film, anti-fungal creams, dyes, pigments and acidity regulator food additive E260. Current production is over eleven million tonnes per year.

Batyric Acid (BTA)

BTA is a fatty acid found in milk, butter and cheeses in varying quantities and is produced by a process called anaerobic fermentation (which is happening inside you as you read this!). Body odour and vomit smell the way they do (yeah, yucky) because of the BTA in them. It was first found in butter and the name comes from the Latin word batyrum, meaning butter. Batyric acid is metabolised by the body and used by mitochondria as an energy source to power cells (I told you it is important).

BTA is produced for industrial use by fermentation and is used to produce salts, and esters (which are compounds such as fats and oils). These are used as additives for perfumes and food (including fishing baits).

Propionic Acid

This acid prohibits the growth of moulds and some bacteria and is mainly used as a preservative for animal feed and food for human consumption (E number E280). It is also used to make pesticides and pharmaceutical products. The esters of propionic acid have fruit-like odours and are used in some solvents and artificial flavourings. On the downside, it is produced by bacteria known as propionibacterium that live on human skin, and is one of the principle causes of acne.

Enzymatic Hydrogen Production

This is a bit cutting edge at the moment but is showing huge potential. In experiments it has been proven that sugar glucose can be used to produce hydrogen using two enzymes; glucose dehydrogenase and hydrogenase.

Glucose is a simple sugar. It is made during photosynthesis from water and carbon dioxide, using energy from sunlight. Glucose circulates in the blood of animals as blood sugar and is the most important source of energy for cellular respiration. The molecular formula for glucose is $C_6H_{12}O_6$.

Research is being carried out to develop this process and it is likely that the result will be a 'hydrogen on demand' system that produces enough hydrogen to run a fuel cell. When that happens, you should be able to generate electricity with a fuel cell using glucose as fuel.

Mining/underground reserves

It was long believed that hydrogen gas reservoirs were extremely rare. It's rare to find them in places where energy companies drill for oil and natural gas, true, but if one knows where to look, they're more common.

One such place are Earth's "cratons," the oldest and most stable parts of the tectonic plates. Some continents have more than one craton, others, like the North American craton, are much larger and so cover most of the continent.

Unlike oil and gas which need thousands of years to form from decomposing organic matter, hydrogen gas is constantly being made underground as water interacts with iron minerals at high pressures and temperatures.

Among these iron minerals is olivine, which, through a chemical reaction called serpentiniZation, steals an oxygen molecule from water percolating down from Earth's surface to transform olivine into serpentinite, and the water into hydrogen gas.

Deposits of olivine are richest in an underground, cratonic feature called a "greenbelt." It's thought that these greenbelts, because of their high concentration of olivine, act as Earth's hydrogen gas engine.

Hydrogen fuel has huge potential to transition off fossil fuels as it's the best currently perceived alternative for diesel or kerosene-based transport such as semi-trucks, jet aircraft, and cargo ships.

Ian Munro, CEO of Helios Aragon, a start-up business pursuing hydrogen in the foothills of the Spanish Pyrenees, estimates his break-even costs will be between 50 and 70 cents per kilo, which is around one tenth of the cost of hydrogen created through electrolysis with solar, wind, geothermal, or other green energies. He feels that this will revolutionize energy production.

Hydrogen Storage Methods

Hydrogen has the highest energy by mass of any fuel. This sounds wonderful but, unfortunately, at normal atmospheric pressure it has a very low density. One kilogram of hydrogen gas, which will run a current fuel cell vehicle for around 100 km, at normal atmospheric pressure and a temperature of 21° C takes up nearly 12 cubic metres. This means it has very low energy by volume, which can only be improved by compression, or cooling it to a liquid or slush state for storage and transportation.

A lot of research is being done at the moment into methods of storing hydrogen inside materials which can then release it as required, doing away with the problems associated with pressurisation and liquefaction.

Pressurised Storage

Currently the most common way of storing hydrogen is under pressure in large underground caverns, typically depleted oil, gas and coal fields, or salt mines. From there it can either be transferred into storage tanks or gas cylinders for transport, or pumped along pressurised pipe lines in a similar fashion to natural and other gases.

At the moment a lot of hydrogen is distributed in steel cylinders at a pressure of 200 bar. (The bar is a unit of pressure defined as 100 kilopascals. One bar is approximately equal to standard atmospheric air pressure at sea level).

To further improve storage capacity, tanks and cylinders made of composite materials have been developed that are lighter than steel and allow pressures of up to 700 bar. This technology has been taken up by hydrogen fuel cell vehicle manufacturers as it means that a 125 litre tank can store 5 kg of hydrogen, which is gives a range of around 500 km for fuel cell cars.

Liquid Storage

Hydrogen starts turning into a liquid at -250° C. To put this in perspective, water freezes at 0° C, the average temperature at the North Pole is -50°C and the lowest temperature possible (absolute zero) is -273.15° C. To store hydrogen in this form requires isolated tanks capable of keeping the temperature below -250° C.

Liquid hydrogen is currently only used in high tech applications, mostly as a fuel for rockets.

Material Based Storage

This is the subject of a lot of research that is beginning to show promising results. There are two main methods, hydrides and LOHCs.

Hydrides

Hydrides are chemical compounds in which hydrogen is combined with other elements. Formally, hydride is known as the negative ion of a hydrogen atom (H^-) and is also called a hydride ion. Because of this negative charge, hydrides have what is called reducing properties. This means that they donate electrons to other elements or compounds. There are three main types of hydride.

Saline (Ionic) Hydrides

Also known as ionic hydrides, these are compounds of hydrogen and active metals. Active (or alkali) metals are elements in Group One of the periodic table and include sodium, lithium and potassium. They react so strongly and quickly with other elements that they do not appear in pure form in the natural environment and require careful handling.

Saline hydrides react instantly with water to produce hydrogen gas and solution. Sodium hydride is more commonly known as caustic soda (or lye) and already has great industrial importance. Because they give off large volumes of hydrogen gas in the reaction with water, it is anticipated that they will be used as light and portable sources of hydrogen.

Metallic (Interstitial) Hydrides

These are formed when hydrogen bonds with transitional metals. These are elements that appear in Groups Two and Three of the periodic table. The better known elements in this category are iron, copper, silver and gold. They are less reactive than the alkali metals, have high density and melting points and are good conductors of electricity.

By combining one or more metals elements to form an alloy, the storage capacity and hydrogen release rate can be modified. Alloys have been traditionally made to give greater strength and resistance to corrosion. Bronze is an alloy of copper and tin, and brass is 70% copper and 30% zinc. There is a lot of research being carried out in this field at the moment.

Covalent Hydrides

Consisting of compounds of hydrogen and non-metals, these are mostly liquids or gases that have low melting and boiling points, except in cases where their properties are modified by hydrogen bonding. One example of a covalent hydride is when hydrogen bonds with chlorine and forms hydrochloric acid (HCl). Others include water (H_2O), ammonia (NH_3) and hydrogen fluoride (HF). Much of the research being done is based around improving the rate of release of the hydrogen.

LOHC (Liquid Organic Hydrogen Carriers)

The storage of hydrogen in this form is seen as a break-through on the path to a hydrogen economy. An organic compound is hydrogenated using a catalyst to store the hydrogen in an easily managed liquid form and then dehydrogenated (again using a catalyst), thus releasing the hydrogen for end use.

LOHCs are similar to diesel and thus can be transported and distributed using the existing infrastructure for oil-based fuels. They also do away with the problem of hydrogen evaporation and can be stored for many months without any losses.

There are already some companies offering this service and the market will only continue to grow.

Practical application and uses of hydrogen

Overview

Hydrogen is already a vital part of our global economy and is set to become even more prevalent as more uses and better production and storage techniques are developed.

Because it is lighter than air, it is used as a lifting agent in weather balloons and airships. Liquid hydrogen combined with liquid oxygen is the most common form of rocket fuel. It is used in fuel cells to produce electricity and can be used as a fuel for internal combustion and jet engines.

Currently, the most common use of hydrogen is in chemical processes and reactions. It is an important element in the process of breaking down crude oil into fuel oil, creating ammonia which is then used to make fertilizer, and as a hydrogenating agent in the manufacture of baked products and biscuits.

The chemical industry uses hydrogen to produce hydrochloric acids, methyl alcohol (for inks, varnishes and paints), hydrogen peroxide (a very important compound for medical use – most First Aid kits contain it) and as part of the process for purifying tungsten.

Welding companies use hydrogen in welding torches that can melt steel, including Atomic Hydrogen Welding (AHW).

Hydrogen fuel cells generate electricity from oxygen and hydrogen. They have a multitude of uses depending on type and are used in weather stations, spacecraft, submarines and, more recently, motor vehicles.

Most spacecraft (including the ISS and Hubble Space Telescope) use re-chargeable nickel-hydrogen batteries because of their long life (they can go through over 20,000 charge cycles).

Industrial use

Hydrogen is used in a range of industries including chemical production, metal refining, electronics manufacturing, food processing and oil refining.

<u>Hydrocracking</u>

This technique is an important source of diesel, jet fuel and other refined fuels.

A hydrocracking unit (called a hydrocracker) takes gas oil and cracks the heavy molecules into distillates (such as jet fuel, kerosene and diesel), and gasoline. There are two main reactions; catalytic cracking of heavy hydrocarbons into lighter unsaturated hydrocarbons, and the saturation of the newly formed hydrocarbons with hydrogen.

Hydrocarbons are organic compounds made up entirely of hydrogen and carbon. The majority of hydrocarbons found on Earth occur naturally in crude oil. The most important thing about hydrocarbons is that they like bonding together to form more complex molecules.

Unsaturated hydrocarbons are simpler hydrocarbon molecules that have the ability to absorb more hydrogen atoms. This means that they are not filled up to their limit (saturated) with hydrogen.

The hydrocracker upgrades low quality heavy oils into high quality refined fuels. Hydrogen gas is also used to regulate the temperature of the hydrocracking process.

Hydro treating is a process whereby hydrogen is used to remove impurities. It can remove up to 90% of contaminants such as nitrogen, sulphur, oxygen and metals from liquid petroleum. Without this process, catalytic converters (the emission control devices fitted to all modern Internal Combustion Engine (ICE) vehicles) would not work.

So, in order to produce the polluting fuels that we currently use in ICEs and jet engines, we have to use hydrogen. I find this very disappointing as hydrogen can actually be used to fuel ICEs and jet engines directly (and only emits water and trace amounts of nitrogen oxide from the exhaust), and as a fuel in fuel cells to generate electricity that can be used to power vehicles, industry and homes.

Ammonia Production

The Haber process converts atmospheric nitrogen into ammonia by a reaction with hydrogen using a metal catalyst under high pressure and temperature.

Ammonia is a compound of nitrogen and hydrogen and has the chemical formula NH_3. It is used as a building block for the synthesis of many pharmaceuticals and is also present in diluted form in many conventional cleaning products. Your kidneys secrete ammonia to neutralise excess acid. Carpets and clothing fabrics rely on ammonia as part of the finishing process but the most important use of ammonia is in in the manufacture of fertilisers.

The Haber process was first demonstrated in the summer of 1909 by the German scientist Fritz Haber and his assistant Robert Le Rossignol. It was quickly purchased by the German chemical company BASF and they assigned the German chemist and engineer Carl Bosch the task of scaling up the machinery for industrial level production. Demand for nitrates and ammonia for use as fertilisers and feed stocks had been steadily increasing during the 19th century. The main source was from the mining of nitrate deposits, and the reserves were rapidly diminishing. The Haber process (sometimes also called the Haber-Bosch process) now produces in excess of 450 million tonnes of nitrogen fertiliser per year. In combination with pesticides, these fertilisers have quadrupled the productivity of agricultural land. Today nearly 80% of the nitrogen found in human tissue originated from the Haber process. Both Haber and Bosch were awarded Nobel prizes for their work in this field.

The Haber Process

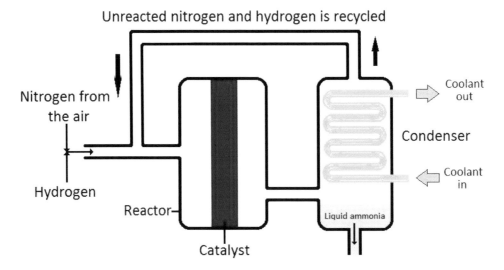

Hydrogenation and dehydrogenation

Hydrogenation is a chemical process based upon the chemical reaction that occurs when a compound or element has hydrogen added to it (by bonding pairs of hydrogen atoms to a molecule). The process requires a catalyst, most commonly nickel. Hydrocracking and the Haber process are examples of processes that make use of this reaction, but it is more widely known as a way of converting unsaturated vegetable oils into saturated fats, similar to animal fats such as lard.

The resulting saturated oil fats are generally less expensive to produce than saturated animal fats and have a longer shelf life. This makes them ideal for use as shortening in most commercial baked goods. Solid or semi-solid fats are preferred for baking because the way they mix with flour produces a better texture in the end product.

The hydrogenation process can be accurately controlled to produce a wide range of consistencies. The major disadvantage of this is that by stopping the process before it is fully complete, some molecules are left in a transitionary condition. These are called trans-fats.

Trans-fats can occur naturally. They are irregularly shaped and the human body cannot process them in the way it can process fully saturated or unsaturated fats. This can cause health problems and, following legislation of varying degrees around the World, they are being phased out. Partially hydrogenated fats can be replaced by blending different oils, for example adding palm oil to unsaturated rich oils or even partially hydrogenated oils.

Fuel Cells

Although fuel cells have been in use for quite a while now, they are becoming more popular due to the ever improving technologies and processes being developed, and the fact that they produce no CO_2. Solar panels (discussed later on) are also in the same position and I believe that these two renewable technologies are the most likely to become mainstream.

Now, let's be clear about power generation. Combustion processes like gas turbines and internal combustion engines burn fuel and use the pressure generated by the subsequent expansion of gases to do mechanical work (pushing pistons up and down in car engines for example), which is used to produce either mechanical forces or generate electricity. The by-product, or waste gases, are pollutants such as carbon dioxide, and heat.

Batteries convert chemical energy into electrical energy. They have a fixed life span and then have to be discarded (which creates a problem because they contain toxic chemicals and acids). Re-chargeable batteries have become an everyday feature of modern life but even so, they need to be recharged using electricity generated by some other method. They also generate heat. Boeing's 787 Dreamliner was grounded for three months in 2013 due to problems including fires caused by the lithium-ion batteries originally used, and over- heating is not uncommon in mobile phones and other modern gadgets.

Fuel cells are devices that generate electricity through electrochemical processes instead of by combustion. They operate quietly and efficiently and produce virtually no pollutants. Fuel cells require a continuous source of fuel and oxygen or air to sustain the chemical reaction, and produce electricity for as long as fuel and air are fed into them. The by-products (emissions or waste gases) are water, heat and, in some cases, trace amounts of nitrous oxide (NO_2), also known as laughing gas.

There are various types of fuel cell, but they all work on the same basic principle. Each individual cell has two electrodes: one positive (cathode) and one negative (anode). The reactions that produce electricity occur at the electrodes. The cell also requires an electrolyte, which can be solid or liquid. The electrolyte is a substance that will only allow the desired type of ion to pass between the electrodes. It is vitally important because if other substances or free electrons reach the electrodes, the reaction is disrupted and the efficiency of the cell decreases.

The anode and cathode contain catalysts that make the fuel oxidise. This creates positively charged protons (hydrogen ions) and negatively charged electrons. The protons are drawn through the electrolyte and the electrons move from the anode to the cathode through an external circuit, which produces Direct Current (DC) electricity. At the cathode the protons, electrons and oxygen react to form water.

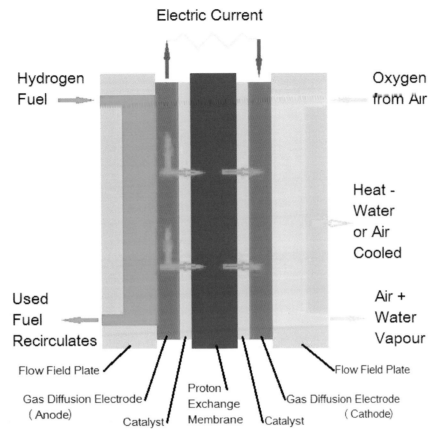

Proton Exchange Membrane Fuel Cell

The individual fuel cells are stacked together to form a fuel stack. There are seven distinct types of fuel cell that each have their own advantages and drawbacks. Here is a list followed by a more detailed look at how they work and what they are most suitable for.

Polymer Exchange Membrane Fuel Cell (PEMFC)

Solid Oxide Fuel Cell (SOFC)

Alkaline Fuel Cell (AFC)

Molten-Carbonate Fuel Cell (MCFC)

Phosphoric-Acid Fuel Cell (PAFC)

Direct-Methanol Fuel Cell (DMFC)

Microbial Fuel Cell (MFC)

Polymer Exchange Membrane Fuel Cell (PEMFC)

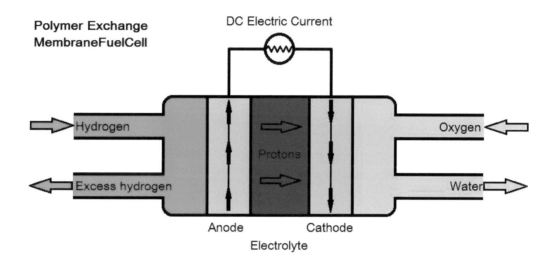

This uses a water based acidic polymer membrane as the electrolyte, and platinum based electrodes. Because of the properties of the electrolyte and precious metals in the electrodes, this type of fuel cell needs pure hydrogen to operate. This means that there has to be less than fifty parts per million of contaminants such as carbon monoxide (CO). The oxygen can be either purified or extracted directly from the air at the electrode.

PEMFCs generate a relatively low level of heat (less than 100° C) and can be used as an 'on demand' system, which increases the overall efficiency of the system.

By changing the electrolyte from water based to a mineral acid based solution you can create a High Temperature PEMFC. These variants can operate at higher temperatures (up to 200° C), which overcomes some of the limitations of fuel purity – they can tolerate impurities of up 5% CO by volume – and are less complex. Drawbacks include lower power density and the fact that they need to be heated before use, whereas standard PEMFCs can be used from cold.

The average overall efficiency ratings are between 40% and 60% (The efficiency rating is the amount of fuel that is converted to useable power). PEMFCs are currently the leading technology for powering light duty vehicles (cars, motorbikes, vans, buses, etc.).

Solid Oxide Fuel Cell (SOFC)

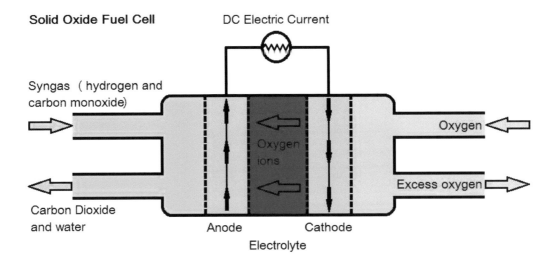

Solid Oxide Fuel Cells use a solid ceramic based electrolyte instead of a liquid or membrane. They operate at very high temperatures, typically 800° C to 1,000° C. This high temperature means that fuels can be reformed within the fuel cell system, which enables them to use a wide variety of hydrocarbon fuels including coal gas. Another advantage is that the reaction rate increases to the point where metal catalysts are not needed for the electrodes.

Disadvantages are that they take longer to start up and reach operating temperature, must be made from robust heat-resistant materials, and require shielding to prevent heat loss.

Although the overall efficiency of SOFCs is around 60%, this can be increased to 80% by harnessing the energy from the heat they produce.

SOFCs can be used for large and small stationary power generation, from 100kw off grid generators to Combined Heat and Power (CHP) systems for use in domestic situations. They are also being developed for use as small portable chargers.

Alkaline Fuel Cell (AFC)

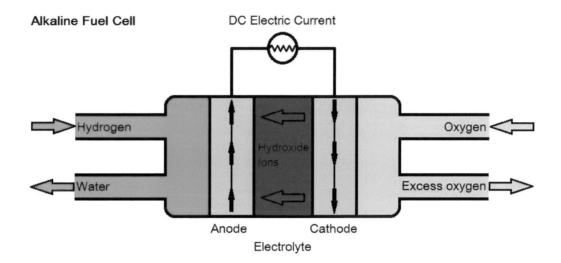

AFCs were one of the first fuel cells to be developed and NASA used them to produce both electricity and water aboard spacecraft including the Space Shuttle.

As the name suggests, AFCs use an alkaline electrolyte and require pure hydrogen for fuel. Early cells had an operating temperature of over 100° C but the latest ones operate at around 70°C. This means that the traditional platinum (used because it outperforms other catalysts) can be replaced by a variety of non-precious (and therefore cheaper) catalysts, usually nickel. They have a high rate of chemical reaction and therefore the fuel to electricity conversion efficiency can be as high as 60%.

AFCs are currently the cheapest cells to manufacture but are prone to carbon dioxide poisoning (a build-up of CO_2 around the anode and cathode) that reduces their efficiency. The latest technical advances in fuel cell research have overcome this problem and AFCs are now being developed for commercial use. This should lead to wider use in the transport industry. Previously they have been used in space vehicles and underwater exploration vessels.

Molten Carbonate Fuel Cell (MCFC)

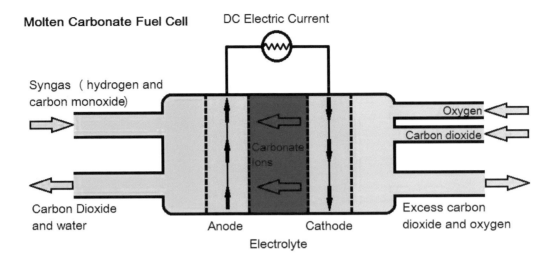

These use a molten carbonate salt solution in a porous ceramic matrix as the electrolyte. They operate at high temperatures, typically 650° C, which gives them the same advantages as SOFCs and enables them to achieve similar levels of efficiency.

Early MCFCs suffered from high temperature corrosion and the corrosive nature of the electrolyte but the latest versions have this under control and thus have a practical lifetime. MCFCs can be used in a stationary facility to generate large amounts of electricity in the order of megawatts. They are also practical for use in CHP systems.

Phosphoric Acid Fuel Cell (PAFC)

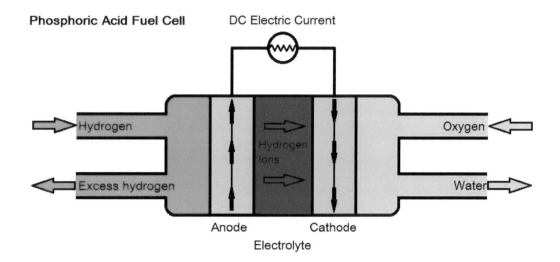

The electrodes in this type of fuel cell are made of a carbon core with a light coating of platinum as the catalyst. The electrolyte is phosphoric acid (hence the name) that is contained in a silicon carbide structure. Most fuel cell units sold before 2001 (the year, not the film) used this technology. As standard they are the least energy efficient of all the fuel cell types but their operating temperature of about 180° C means that the heat can be harnessed for co-generation, which can raise their overall efficiency as high as 80%.

PAFCs are used for stationary power generation in the 100 to 400kw range (mostly for commercial purposes). They are also being developed for use in large vehicles such as trucks and buses.

Direct Methanol Fuel Cell (DMFC)

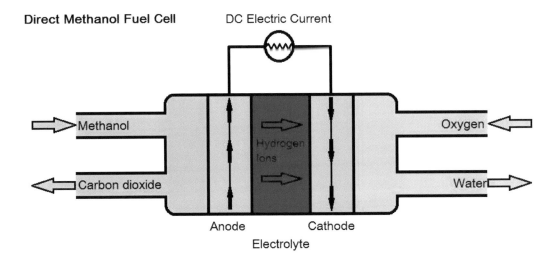

These were developed in the 1990's by researchers in several U.S. institutions including NASA and the Jet Propulsion Laboratory (JPL). They use a polymer membrane electrolyte and are similar to PEMFCs but use a platinum-ruthenium (a chemical element, symbol Ru, atomic number 44) catalyst on the anode. This enables it to draw hydrogen directly from liquid methanol.

Methanol is inexpensive and can be easily transported and stored. It has a relatively high energy density and is useable in a liquid reservoir (that can be topped up) or a cartridge (that is replaced when empty).

DMFCs operate in a temperature range from 60° C to 130° C and have an overall efficiency of about 30%. They are currently used in applications that have modest power requirements such as chargers or portable power packs. They are also used to power materials handling vehicles such as fork lift trucks. Conventionally, fork lift trucks run on rechargeable batteries. Switching to fuel cells saves time in refuelling (minutes instead of hours for recharging) and frees up the space required for the battery storage and infrastructure.

Microbial Fuel Cell (MFC)

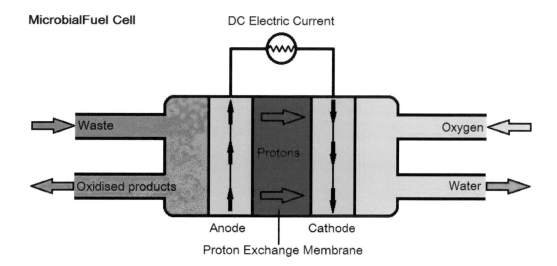

Microbial fuel cells are different from all the others in that they use living bio-catalysts instead of chemical catalysts. Although the idea has been around for about a hundred years, it was not until 1999 that a scientific breakthrough enabled practical microbial fuel cells to be made.

MFCs use the process of cellular respiration. This is what cells use to convert nutrients into a substance called Adenosine Triphosphate (ATP), which is the fuel used for cellular activity. Bacterial respiration is the oxidisation and reduction of organic molecules and this process naturally moves lots of electrons around.

The big breakthrough in 1999 involved the use of a recently discovered type of micro-organism called exoelectrogens. These are a special class of bacteria that will stick to the surface of an anode coated with a specialised protein and then transfer electrons to it.

Organic fuel enters the anode chamber and the bacteria get to work converting it to ATP. The by-products of this process are protons, electrons and CO_2. The electrons are accepted into the anode while the protons pass through a proton exchange membrane into the cathode chamber. An oxidising agent recombines the electrons from the cathode with hydrogen and produces water.

The efficiency of MFCs can be up to 50% and this is expected to improve as further research is completed. They typically operate at between 20° C and 40° C. The conditions of a waste water treatment plant are ideal for the bacteria that can be used. The energy produced offsets the cost of running the plant and the bacteria eat up around 80% of the sludge present in waste water. The latest developments produce clean water, electricity and methane, which can be used as a clean fuel.

Five general advantages of fuel cells over conventional power sources

1) Efficiency

Energy efficiency is basically the amount of useable energy produced using a specific unit of fuel. Below is a list of the major methods of energy production currently in use, with their efficiency levels.

Fuel Cells:	Between 40% and 80% depending on type
Diesel ICE:	Up to 20% for modern engines (typically 15% to 18%)
Petrol ICE:	Up to 20% for modern engines (typically 12% to 16%)
Wind Power:	40% to 45% in optimal conditions. Usually a lot less
Nuclear Conversion:	Around 38% for the thermal cycle but only 0.7% for overall energy conversion.
Coal Fired:	Typically 35% to 40%
Natural Gas:	Typically 35% to 40% but the latest technology can achieve up to 60%.
Hydro-electric:	85% to 90%.
Geo-thermal:	Around 35%

2) Fuel cells, hydro-electric, geothermal, solar, wind, wave and tidal power are the only methods of generating power that don't produce pollutants and dangerous by- products.

3) Fuel cells require just a little bit of simple maintenance due to the fact that they have very few moving parts.

4) There is little or no noise pollution when using fuel cells. In fact most of them produce no audible noise at all.

5) Using stationary fuel cells to generate power at the point of use would enable a decentralised power grid that is more stable and allow power on demand for everyone.

Recommended further reading

How Does Hydrogen Fuel Cells Work - Hydrogen Fuel News

Home | Hydrogen Fuel Cell Partnership (h2fcp.org)

The Centre for Fuel Cell and Hydrogen Research, Chemical Engineering, University of Birmingham - University of Birmingham

First of its Kind Discovery in Mali: Vast Reservoirs of Clean Hydrogen Gas - Hydrogen Central (hydrogen-central.com)

Helios Aragon | Solutions to the green hydrogen deficit. (helios-aragon.com)

Where are we now?

The world is taking the first proper steps towards a sustainable hydrogen economy that will bring huge benefits to mankind. As we progress further along this path and combine it with other forms of green and renewable energy, the amount of pollution will drop dramatically and the quality of the air, particularly in large cities, will improve, bringing with it a decrease in respiratory and other health problems.

The main emphasis at the moment is on introducing hydrogen powered vehicles but it is also being used to generate electricity as an alternative to diesel generators. First of all though, the hydrogen has to be produced.

Generating or manufacturing hydrogen fuel

As the concept and reality of using hydrogen as a fuel are gaining momentum, there are many projects taking place in universities and research centres looking into ways of producing hydrogen cleanly and cheaply. Here are some of the most promising ones.

USC – University of Southern California

I have singled out this university because of its extremely illustrious history and importance in developing technologies.

USC is a private, non-profit and non-sectarian research university founded in 1880. The main campus is in Los Angeles and it is California's oldest private research university. It quickly became an engine for economic activity and contributes billions of dollars into the economy of Los Angeles County every year. It also has links with many top research facilities worldwide.

Chemists at the USC Loker Hydrocarbon Research Institute have found a way of producing hydrogen by recycling carbon dioxide through a process whereby they are able to dehydrogenate formic acid. The breakthrough over other similar research is that the catalyst they have developed releases hydrogen without producing any polluting by-products.

Formic acid occurs naturally in the venom of bee and ant stings and through the decay of vegetation. The fact that it can also be made by reacting carbon dioxide with hydrogen makes it an appealing option for scientists as a carrier and storage reservoir for hydrogen.

The breakthrough made at USC Loker is the development of a new iridium based catalyst to release the hydrogen. This catalyst has some major advantages: it doesn't degenerate when exposed to air and is almost inexhaustibly re-usable.

Professor Travis J. Williams, lead author of the study, is on record as stating, 'We think it may never die. We've demonstrated it through 2.2 million turnovers over several months. At that point, the catalyst was still alive but we were bored with the experiment. This would be a big money save for anyone who wants to generate and store hydrogen as a fuel.'

They used 'neat' formic acid (with no added water or solvent) because when you release energy from a liquid fuel you want to have as little unnecessary mass as possible. Think of it this way: If there is twice as much fuel in a gallon of liquid then your vehicle will go twice as far on each tank of fuel.

The concept is basically the reverse of combustion. It takes the CO2 released as a result of burning carbon based fuels and turns it into a useable non-polluting fuel.

This development will essentially lead to a 'hydrogen on demand' system that produces hydrogen directly for a linked fuel cell to generate electricity, or straight into an internal combustion engine designed to run on hydrogen. So far all the indications are that this process is inherently both safe and sustainable.

Large Scale Manufacturing

There are many companies around the world that manufacture hydrogen already. More research and resources are being put into supplying hydrogen (and other fuels for use in fuel cells) all the time. The major downside to this approach is that the fuel has to be transported to the point of use, usually by pipeline or in cylinders.

On Site Generation

There is a lot of interest in producing hydrogen on site for re-fuelling fuel cell vehicles (FCEVs) and the number of companies able to provide this equipment is increasing rapidly.

One important pan-European project called Hydrogen for Innovative Vehicles (HyFive) ran from 2014 to 2018 to facilitate rolling out FCEVs and the required re-fuelling infrastructure. In May 2016 the first London HyFive hydrogen re-fuelling station opened in Teddington. It uses ITM Power H Fuel hydrogen generating stations that run on electricity to generate hydrogen by electrolysis, which is then compressed and stored to be dispensed on demand. It needs a water supply as well as the electricity, and if the electricity comes from a renewable or green energy source then the entire process, including running the FCEVs, is carbon free. Currently, their highest capacity unit produces enough hydrogen fuel for 94 vehicle refills every 24 hours.

Hydrogen fuel vehicles

There are currently three FCEV manufacturers with vehicles available to own or lease. These are Hyundai, Toyota and Honda. Most of the major car manufacturers will be introducing FCEVs in the next few years including BMW, the Audi group (including Volkswagen) and Mercedes Benz. There are also some start-up companies preparing to join in, such as Riversimple. At the same time, they are starting to scale back on the production of diesel powered vehicles. Using hydrogen as a fuel for internal combustion engines is also gradually gaining momentum.

How do fuel cell cars work?

So, we know what fuel cells are, the different types and how they work. Hydrogen fuel cell technology is an area of the automotive industry that is becoming increasingly important, with manufacturers committing to developing FCEVs.

The most common fuel cell for powering vehicles at the moment is PEM. Many individual cells are combined into a fuel cell stack to generate the required voltage. As we have already discussed, the by product is water.

Basically, there is a hydrogen fuel tank, a fuel cell stack, oxygen intake, electric motors and a storage battery. Hydrogen from the tank and oxygen from the air flow into the fuel stack, which produces electricity to drive the motors. Usually there is a motor for each wheel, so the car is four wheel drive. Most will also use regenerative braking to put electricity back into the system as well. The by-product of the fuel cells is water, which is released into the environment. Currently FCEVs have a range of 300 to 350 miles per tank of hydrogen and the re-fuelling time is three to five minutes, comparable to current petrol and diesel vehicles. Because it is also a mobile electricity generator, the vehicle can be fitted with a power output socket and could be used to run electrical appliances of all descriptions. So, you could go camping and have a TV, cooker, lights, etc. powered by your car.

Other vehicles

Buses

There are many start-up programs around the World using fuel cells to power buses (FCEB). The first FCEB was demonstrated in 1993. FCEB technology reduces emissions and will achieve better fuel economy than conventional diesel powered buses.

According to the U.S.A.'s Federal Transit Administration (FTA) who have been running assessments and development programs on FCEB for a number of years under their American Fuel Cell Bus Project, every single FCEB has the potential to reduce the amount of carbon dioxide released into the atmosphere by 100 tons per year and eliminate the need for 9,000 gallons of fuel every year over the life cycle of the vehicle. This translates into savings of $37,000 per vehicle annually over conventional diesel fuel buses. They are also quieter and smoother, especially when idling or stationary. As my daughter said when I told her about them, you would be able to rest your head against the window without having your teeth rattled loose when the vehicle is stationary.

In Europe there was the CHIC project (Clean Hydrogen in European Cities). CHIC was a public-private partnership and received substantial funding from the Fuel Cell & Hydrogen Joint Undertaking (FCH JU). The project was completed in December 2016. CHIC operated 54 hydrogen fuel cell (H2FC) buses and 4 hydrogen powered internal combustion engine buses in 9 cities in Europe and Canada. The buses were delivered by 5 different bus manufacturers and had fuel cells from two different suppliers. CHIC had many partners, including Transport for London, Daimler, Total, Shell and The University of Stuttgart. Demonstration sites included Berlin, Cologne, Hamburg, Milan and London, with related projects in Rotterdam, London, Antwerp, Cherbourg, Rome, Aberdeen, San Remo and Stuttgart.

In late 2016 London Mayor Sadiq Khan is quoted as saying, "I want London to become a world leader in hydrogen and electric bus technology. I'm implementing hard-hitting measures to clean up London's toxic air and it's great that more cities are getting on board to phase out the procurement of pure diesel buses, which sends a clear signal that only the cleanest technologies are wanted in our cities."

The governments of eleven big cities (Amsterdam, Cape Town, Copenhagen, Los Angeles, New York, San Francisco and others) responded to Khan's call and the number of fuel cell buses in use Europe-wide is set to increase to around 300 to 400 in the next few years.

The mayor of Copenhagen, Frank Jensen, added: "In Copenhagen, all new buses will be based on zero-emission and low-noise technologies from 2019."

Denmark's capital has long put sustainability at the top of its agenda, as have many of its inhabitants. Lots more initiatives are taking place this year as the city continues its bid to become carbon-neutral by 2025.

All buses are changing from diesel to electric, while more and more road surfaces are devoted to cycling.

A significant landmark is Copenhill, which turns waste into energy to supply tens of thousands of homes and businesses. It is covered by a year-round artificial ski and snowboard slope, one of the longest in the world and an exciting novelty for both residents and visitors. Hikers are welcome in a new green space up top.

Rehman Chishti, Conservative MP for Gillingham and Rainham in the UK is a big supporter of the hydrogen economy and stated the following: 'In road transport, battery electric vehicles (BEVs) are rightly seen as a key pillar of decarbonisation. Hydrogen propulsion also offers great potential, especially in heavier and longer distance vehicle segments, due its longer range and fast refuelling. Hydrogen fuel cell electric vehicles (FCEVs) are expected to play a major role in the medium to long term. Hydrogen internal combustion engines (hydrogen engines) present an alternative path towards zero emissions that has many benefits in the shorter term.'

A company called Wrightbus has supplied nearly all of the 100 or so hydrogen buses in service in the UK today and Chairman Jo Bamford has said he plans to put 3,000 hydrogen buses on the streets by 2024.

Meanwhile, Japan is one of the world leaders in the move towards a hydrogen economy. Currently there are eighty-five hydrogen-powered buses in use in Japan, seventy owned by the city of Tokyo and fifteen in the private sector. In partnership with the Fukushima region, which produces hydrogen in the largest factory in the world to supply vehicle batteries, the city of Tokyo intends to equip more than three hundred buses by 2030.

Hydrail

Short for Hydrogen Rail, this is a generic term referring to rail based transportation that uses hydrogen as fuel, including vehicles such as trams as well as railway locomotives. Fuel cell trains will play a key role in the transition to a zero emission economy. Hydrogen powered trains are poised to disrupt the rail industry as a cost-effective, high performing, zero-emission alternative to diesel.

Hydrail uses a hybrid configuration of hydrogen fuel cells, batteries and electric traction motors. The fuel source is hydrogen. The fuel cells convert the hydrogen into electricity, which feeds the batteries to provide a stable power source for the traction motors. The train's brakes recharge the batteries, which store any excess energy for later use, and further contribute to fuel efficiency. They take less than 20 minutes to refuel, and can operate for more than 18 hours between refuelling. Using Hydrail removes the need for overhead and third rail electricity infrastructure and power substations.

Germany has taken a step forward by getting the world's first hydrogen-powered trains up and running. Five Coradia iLint trains started carrying passengers in August 2022, and nine more will replace the diesel trains currently running on a route in Bremervörde, Lower Saxony in early 2023. The driving force behind this and other hydrogen fuelled transportation is a company called Cummins, in partnership with Alstom.

There are plans for the trains to expand to other parts of Germany and Europe: Frankfurt has ordered twenty-seven trains for its metropolitan area, France intends to deploy twelve, and the northern Lombardy region of Italy is starting with six. The U.K. has also run a test vehicle and is preparing to start using production models in the near future.

Japan, China and the U.S.A. all have their own Hydrail projects with fuel cell locomotives already developed and tested.

Internal Combustion

Apparently, most modern internal combustion engines (ICEs) can be converted to run on hydrogen, and even be switchable from hydrogen back to petrol if there is no hydrogen refuelling available. It is mostly about getting the engine management correct and installing the hydrogen fuel tank.

An ICE designed specifically for hydrogen is much more efficient though, and there is less chance of it going wrong. Research shows that the power output of a direct injected hydrogen engine is about 20% higher than an equivalent petrol engine. Direct injection engines put out about 40% more power than those using a carburettor.

Hydrogen hybrid ICEs, either using hydrogen from a refillable tank or an on demand system using electrolysis, combine the hydrogen with the petrol or diesel fuel/air mixture before it enters the engine.

As much as 60% of the fuel in a standard ICE is lost as heat. Introducing hydrogen makes a cleaner fuel burn that reduces engine and exhaust gas temperatures and improves power and efficiency.

Aston Martin entered an HH powered Rapide S in the 2010 Nurburgring 24 hours race and finished second in the SP8 class. It was the first car to compete in the race using hydrogen fuel. While the vehicle did not rely solely on hydrogen fuel, it is considered the first hydrogen-powered model to complete a single lap of the Nurburgring in an official race. Alset Global, the company that developed the HH technology engine, won the Powertrain of The Year award at the Professional Motorsport World Expo held in Cologne in 2013.

According to JCB, one of the world's top three manufacturers of construction equipment, batteries are not practical for equipment with high power demands, or for machines that work in remote locations such as backhoe loaders and large excavators. Batteries would weigh too much, cost too much, and there would not be enough time to charge them even if an on-site charging infrastructure was in place.

They have developed the first hydrogen motor in the industry. Using and adapting established engine technology with readily available components, hydrogen is combusted and power is delivered in exactly the same way as a diesel engine. Hydrogen is stored as gas under high pressure, which is stepped down to a lower pressure when injected into the engine. JCB's unique combustion process involves lowering the pressure and temperature within the engine to give a clean burn and a zero-emission product while delivering exactly the same power and torque as a diesel engine. What's more, the technology is far less complicated than hydrogen fuel-cell technology. Nothing but steam is emitted from the tailpipe, meaning that there is zero CO_2 emission at point of use.

Having developed dual fuel engines which operate on 85% hydrogen gas and 15% conventional fuel that make the owner or user less dependent on fossil fuels, reduces CO_2 emissions by as much as 85% and can also be used flexibly when hydrogen is (temporarily) not available, BEH$_2$YDRO has been working hard on an additional, completely new range of 100% hydrogen engines. At the World Hydrogen Summit held in Rotterdam in early May 2022, BEH$_2$YDRO launched its 100% hydrogen engines for heavy duty applications. These innovative, zero emission engines are ready to use and will allow the industry to further engage in 100% sustainable and environmentally friendly development. BEH$_2$YDRO is a joint venture between ABC and CMB.TECH. Their website gives the following reasons why they are developing hydrogen technologies as opposed to batteries:

They emit no harmful emissions such as carbon (CO_2), nitrogen (NOx), sulphur oxides (SOx) or soot particles. In addition, they are non-toxic and:

- No use is made of scarce raw materials such as lithium, zinc, cobalt, platinum and rare earth minerals.
- Avoid environmental damage caused by the extraction of scarce raw materials.
- They do not cause any damage to human health.
- The by-products are completely safe (the emissions consist of water vapour and air sucked in).

They also have a long service life, thanks to an optimised air/gas mixture that reduces component heat-up.

A common complaint when refuelling on site with diesel is that it can often get spilled, resulting in a dirty smelly mess with associated safety and environmental concerns. Hydrogen refuelling on the other hand is a safe, clean and a fully sealed process with no chance of fuel spillage.

Porsche has developed a hydrogen engine prototype for luxury sports cars that will match a V8, 4.4 litre ICE while also reducing fuel consumption and maintaining emissions equivalent to ambient air.

Hal Reisinger, the CEO of engine specialists Cosworth recently told Autosport magazine that internal combustion engines can be converted to run on hydrogen. The fuel injectors need to be changed and the ECU remapped. This has the potential to make an entire infrastructure of engines available. "It's much easier to establish a hydrogen infrastructure than an electric infrastructure. It's the present," he said in response to a question about how far away a hydrogen powered ICE is for Cosworth.

Hydrogen is competitive with traditional fuels in terms of range and refuelling rates. Toyota have raced a Corolla powered by a hydrogen ICE in the Fuji 24 Hours in 2022 and demonstrated a hydrogen ICE GR Yaris on a stage of the Ypres rally.

What I would really like to see is motorsport switching to hydrogen ICEs as opposed to fuel cells or E-fuels. The potential is enormous and would drive the development for consumer use. There would still be the noise associated with motorsport, which is a drawback with the electric racing formulas such as Formula E, and I'm convinced that most petrol-heads would become hydrogen-heads very easily. Imagine Formula 1 using loud powerful V8, 10 or 12 cylinder hydrogen ICE powertrains. I'm sure that would generate much more interest than the current hybrid technology.

Then there is the potential for swapping out standard internal combustion engines for hydrogen ones. The fuel system would also need to be changed but in most cases this would enable people to keep their existing vehicles at a lower cost than purchasing a new vehicle. The classic and vintage car market could benefit hugely from this. All it really needs to catch on is for the fuel networks to be established and publicised.

A company called Layer and U.S. start-up Croft have joined forces to design and market several new products that will make it possible to retrofit existing vehicles to run on hydrogen.

"Batteries are great to decarbonise smaller passenger vehicles that get used in gentle, predictable ways with access to good charging infrastructure," said Benjamin Hubert, founder of the Layer creative agency. "However, lots of mobility applications don't match that description, and there, we need a power source that is denser than batteries and has fewer dependencies on infrastructure."

"Hydrogen stores significantly more energy in less space and with less weight than batteries, and it's much easier to use hydrogen in environments with weak grids or where charging otherwise isn't available," added Hubert. He explained that, at least in the near future, hydrogen cars and trucks could provide a complementary zero-emission vehicle alternative to electric vehicle technology, as opposed to being competition.

"It's a great complementary solution to batteries, and as with all things, it's important to pick the right tool for the right job," he stated in a recent Dezeen article.

<u>Aeroplanes</u>

According to the Guinness Book of Records website, the first manned hydrogen powered aircraft was a two seat Dimona motor glider using fuel cells to power an electric motor which drove the propeller. The flights took place at an airfield in Ocana, southern Spain between February and March 2008.

On September 29th 2016, a twin cabin plane called HY4 was successfully tested by German engineers at Stuttgart airport. The flight test of the World's first four seater hydrogen powered aircraft lasted ten minutes. Developed by aircraft maker Pipistrel, fuel cell specialist Hydrogenics, the University of Ulm and the German Aerospace Centre DLR, the HY4 uses emission-free hybrid fuel cells to generate electricity from hydrogen, which powers the aircraft during flight. Batteries are used for take-off and landing. It has a range of up to 1,500km (930 miles) at a cruising speed of 165 kph (102 mph).

Rolls Royce and Easyjet have confirmed the first successful tests of a modern jet aero engine running on hydrogen. The test took place at an outdoor test facility at MoD Boscombe Down in the U.K. using a converted Rolls-Royce AE 2100-A regional aircraft engine. Green hydrogen for the tests was supplied by EMEC (European Marine Energy Centre), generated using renewable energy at their hydrogen production and tidal test facility on Eday in the Orkney Islands, U.K. Inspired by the global, UN-backed Race to Zero campaign that both companies have signed up to, committing to achieve net zero carbon emissions by 2050, the partnership plans a series of further rig tests leading up to a full-scale ground test of a Rolls-Royce Pearl 15 jet engine, stating 'hydrogen offers great possibilities for a range of aircraft, including Easyjet-sized aircraft'.

In November 2022 Airbus revealed that it is developing a hydrogen-powered fuel cell engine. The propulsion system is being considered as one of the potential solutions to equip its zero-emission aircraft that will enter service by 2035.

In December 2022 Pratt & Whitney and Rolls-Royce announced a joint research project with Virginia Tech.

There are many other projects being run around the world and research into manufacturing, storage and corrosion resistant materials is beginning to enable the safe use of green hydrogen as the only realistic alternative to fossil fuels.

A company called Reaction Engines has been working on a new type of engine called SABRE - Synergetic Air Breathing Rocket Engine – which is a new class of engine for propelling both high speed aircraft and spacecraft. Part of enabling this ground-breaking technology has been the development of next-generation thermal management technologies that increase performance, efficiency and sustainability. They now work in partnership with many universities and companies that are developing the next generation of transport and energy supply systems.

Maritime and shipping

The general consensus in the maritime industry is that switching to hydrogen power is viable, and there are various projects underway. The major stumbling block at the moment is the same as the other transport systems – cost and availability of fuel, but this is gradually being resolved.

Debuted in 2004, the DeepC is a hydrogen-fuelled Autonomous Underwater Vehicle (AUV), power-assisted by an electric motor that gets its electricity from a fuel cell. The project was funded by the German Federal Ministry for Education and Research.

Bristol Hydrogen Boats and Auriga Energy teamed up to produce the U.K.'s first fuel cell ferry. Called Hydrogenesis, it carries twelve passengers and operates in Bristol between Temple Quay and the SS Great Britain. £225,000 was funded by Bristol City council and the vessel won the Environmental Innovation of The Year award in 2011, before entering service in July 2013. The fuel cells need about ten minutes to recharge, compared with conventional batteries which require about five hours to be fully charged.

Bristol, by the way, has been steadily investing in energy efficiency and renewable energy and won the European Green Capital award in 2015. More details of the winners since its launch in 2010 can be found on the European Commission site.

A key advantage of hydrogen over other fuel alternatives is the relative ease of retrofitting existing ships. With the recent advances in ICEs, the existing diesel engines could be replaced with hydrogen ones. A lot of progress has already been made with hydrogen fuel cells. A recent study by the International Council on Clean Transportation (ICCT) found that 99% of the transpacific voyages made in 2015 could have been powered by hydrogen and fuel cells. More than half of those required only minor fuel capacity changes or just one additional port of call to refuel.

"Maritime is a large contributor of global greenhouse gas emissions and thus a priority sector to decarbonise," said Bart Biebuyck, executive director at FCH2 JU, part of the EU's Horizon 2020 innovation initiative. "Hydrogen and fuel cells have the potential to propel vessels in a zero-emission fashion and various ship types are starting to integrate them."

In Europe, hydrogen is part of the EU's long-term strategy to become carbon neutral by 2050. Fully electric and electric-hybrid ships already operate in several countries, including Norway, and now a new project aims to build a zero-emission prototype vessel running on liquid green hydrogen (LH2).

Liquid Hydrogen – The Fuel of Choice for Space Exploration

All current space launch vehicles use rocket technology. The first stage of NASA's Apollo moon missions used kerosene and liquid oxygen. The second stage used liquid hydrogen instead of kerosene. The main reason for this is that although liquid hydrogen is capable of providing better performance, it is not as dense as kerosene and so requires larger fuel tanks. For every litre of liquid oxygen burned in the Saturn V first stage it used 0.64 litres of kerosene, while the second stage used 3.25 litres of hydrogen.

Today, liquid hydrogen is the signature fuel of the American space program and most other rocket launch systems for their upper stages. The Space Shuttle main engines ran on liquid hydrogen.

SpaceX are now switching from kerosene to methane (CH_4) and liquid oxygen in their Starship's Raptor engines as it burns more cleanly and is far superior for reusable rockets. Other rocket launch systems are making the same move.

Fuel cells are used to provide electricity for human spaceflight. Originally alkaline cells were used, from the Gemini missions through to the Space Shuttle and International Space Station (ISS). Now they are switching to PEMFCs which are becoming more powerful, lighter, safer, simpler to operate and increasingly reliable.

Static Generation

Uninterruptible Power Supply (UPS)

This is an apparatus that provides emergency power when the mains power fails. They are typically used to protect hardware such as computers, data centres, telecommunications equipment, medical equipment and any other electrical equipment where an unexpected power disruption could cause injuries, fatalities, serious business disruption or data loss.

They work by supplying energy stored in batteries, superconductors or flywheels nearly instantaneously in the event of a power failure. They need to be able to supply power for long enough to enable an emergency power system or standby generator to start up, or for the protected equipment to be safely shut down.

A Line-interactive UPS adds a special type of transformer that enables it to automatically filter any power supply fluctuations to provide stable power to the user.

Hydrogen UPS systems are now available that use PEM fuel cells. Upon grid power interruption the fuel cell takes up the load with the help of a storage battery during the first few minutes of operation.

Emergency power system/Backup generator

These are the next stage of a UPS. Most backup generators run using a petrol or diesel generator but these are gradually being replaced by fuel cell generators. Compared to diesel generators, fuel cells can provide savings of up to seventy-five percent, are almost silent when in operation, and the only exhaust is water vapour.

Independent Power Supply

Hydrogen fuel cell generators can be used to power homes and buildings completely independently from a national power grid. When combined with other technologies such as solar panels, you can have a system that generates hydrogen which is then used to refuel both your car and your home power system. There are many projects around the world developing this technology.

The best way to store energy is to generate hydrogen gas via electrolyzers with solar energy and pure water during the day. Electrolyzers produce hydrogen gas by using an electric current to separate water into its composite gases: hydrogen and oxygen.

During the night or when additional energy is required, the hydrogen is used to produce power with a fuel cell. Storage batteries take the very fast and short loads, while the massive permanent energy load is supplied by the hydrogen fuel cell.

Combining different power sources and storage techniques into a cohesive process will revolutionize the power industry, as most domestic applications will be self-sustaining and do away with the need for a centralized power grid. The benefits will be felt the most in poorer countries and remote locations.

Recommended further reading

Hydrogen House Project - Hydrogen House Project: Home

Accelera | Powering the World with Clean Energy (accelerazero.com)

Fuel Cell & Clean Energy Solutions | Ballard Power

Honda Global | Clarity Fuel Cell - Picture Book

Hydrogen Can Move Into Pole Position - Ryze Hydrogen - Hydrogen Central (hydrogen-central.com)

Layer green hydrogen cars Croft system developed - H2 News (hydrogenfuelnews.com)

Other clean energy production methods

Solar Energy

Solar energy is essential for the existence of life on our planet. The Earth orbits the Sun at a distance of 150 million kilometres (93 million miles) and its gravitational pull keeps all of the planets in their orbits. It radiates light and heat (solar energy). Without this energy there would be no winds, clouds or ocean currents and the Earth would be a frozen, dead planet.

Plants use photosynthesis to convert solar energy into chemical energy that enables them to live and grow. It is therefore essential for agriculture – cultivating land, growing crops and raising livestock. It can also be used for drying agricultural produce, which extends its useable lifespan. This is was a major factor in the development of our civilisations.

Early civilisations began using buildings to take advantage of the heating properties of the Sun and it became an important part of their religion and culture. The Romans are generally credited with the invention of greenhouses, which make it possible to grow plants in areas that are not climatically suitable for them, and to cultivate them out of season.

The Greek mathematician, philosopher and inventor Archimedes is alleged to have developed a system of mirrors designed to concentrate sunlight to one point, generating enough heat to set fire to enemy ships. This technique has popped up at various point in history, including the first recognised solar furnace, which was developed by the French chemist Antoine Lavoisier during the early 1770's to further his studies into combustion, and led to huge advances in the field of chemistry.

Solar cookers can convert up to 80% of the solar energy into heat. Conventional box cooker designs can be used to sterilize water and cook meat, vegetables and breads. More advanced parabolic and vacuum tube solar cookers can reach higher temperatures, making them suitable for grilling and boiling water as well. They are most appropriate for use in equatorial regions and are becoming more popular all the time as they do not use fossil fuels and are non-polluting. It is estimated that there are over half a million already in use in India alone.

Solar water heaters were first introduced in the late 1800's but were soon replaced by oil and gas systems. They are becoming popular again, especially in countries including China, Greece and Japan. Australia, Israel and Spain have passed laws that require them to be included in any new construction projects.

There are currently two methods for converting sunlight into electricity: Concentrated Solar Power (CSP) and Photovoltaic (PV).

Concentrated Solar Power

This uses lenses or mirrors and tracking systems to focus a large area of sunlight into a small beam. It collects heat, and uses that to create steam for electricity generation. There are four main types:

Solar Tower

Also known as central receivers, these use flat mirrors to focus sunlight onto a central tower. This heats up a liquid (usually molten salt) which creates steam for use in electricity generation. Solar Towers are more efficient and cost effective than other CSPs and also provide better energy storage capability.

There currently six major solar tower projects in operation. Two are in the U.S.A. - Ivanpah in the Mojave Desert, California, and Crescent Dunes in Tonopah, Nevada. Another two are in Spain –Gemasolar in Seville and PS20 in Sanlucar la Mayor. China has Shouhang Dunhuang, and Israel recently completed the Ashalim Solar Thermal Power Station in the Negev desert.

Solar towers are proving to be a very reliable concentrated solar power technology. Although solar power tower projects are only feasible in areas with enough free land and reliable clear skies, the power produced can be fed into the grid and used for residential and commercial purposes.

Parabolic Trough

As you have probably surmised from the title, this design of CSP uses curved mirrors that concentrate the sun's energy at a central focal point. It works at all levels from very small scale, where it can be used to heat food and liquids, to large scale.

There are around fifty large scale Parabolic Troughs in use around the world, with more under construction. Large scale Parabolic Troughs use a tube (usually a Dewar Tube, which is insulated in a similar fashion to a vacuum flask) that runs along the focal point. The liquid inside it is heated by the reflected sunlight and can be used in a heat engine to generate power or to directly drive machinery. Efficiency ratings for this method are currently comparable to Photovoltaic methods.

Fresnel Reflectors

These are made of thin, flat, mirror strips that concentrate sunlight onto tubes to heat a liquid, which is then used to generate power. Flat mirrors allow for more reflective surface than parabolic reflectors, capture more of the available sunlight and are cheaper to manufacture. Operating and maintaining them is relatively easy, and this is an emerging technology that could potentially replace Solar Towers and Parabolic Troughs because it also has the best land to electricity ratio of any CSP. This is mostly due to the fact that the land beneath it can still be used for agriculture.

Overall, CSP is a promising technology but is limited due to the need for large areas of land and high concentrations of sunlight. Having said that, however, large facilities situated in and around desert areas could be used to add a lot of electricity to national and international power grids.

Photovoltaic (PV)

PV applies to systems that work by converting light into electricity using semiconducting materials.

The photovoltaic effect is a phenomenon that is being continuously studied around the world, especially in the fields of physics, photochemistry and electrochemistry. A typical PV system uses solar panels (made up of small solar cells) that generate electrical power. This method does not generate any pollutants. PV systems have been in use for over fifty years, mostly in specialised applications such as power systems for satellites and spacecraft. In the last twenty years they have been introduced more generally in standalone and grid connected systems. The technology and manufacturing of PV systems is advancing rapidly and over 100 countries now use it in some capacity. It is currently the third largest renewable energy source worldwide, after Hydro Electric and Wind Power.

Okay, I hear you ask, how does PV work? Well, to start with let's find out what the photovoltaic effect is.

It was figured out by a French physicist called Edmund Becquerel in 1839. What he discovered is that some materials produce a small electric current when exposed to sunlight.

Solar cells are small devices that convert sunlight into electricity. They consist of two wafer thin layers of silicon crystal, called P-type and N-Type. The P-type have been heated to make them attract electrons. The N-type is placed on top and has been treated to make it release electrons. When sunlight hits the N-type layer, it excites the electrons and this gives them enough energy to move. The electrons then flow into the P-type layer and generate an electrical current.

Electrical contacts are added to both the top and bottom layers and a protective layer and anti-reflective coating are added to the top. Solar panels produce DC current, so for domestic use they are connected to an inverter that changes the current to AC.

Solar Cell Structure

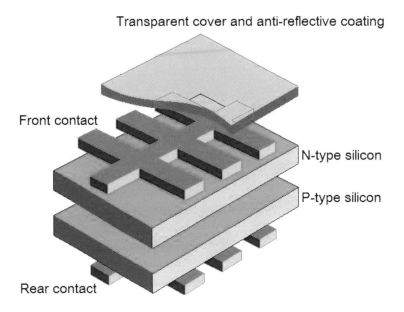

Solar cell process

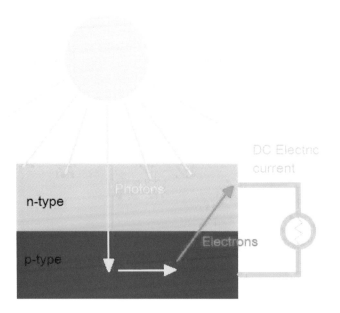

A *solar panel* (or module) is made up of several solar cells connected together and fitted into a frame, which produces a larger amount of power.

A *solar array* is several solar panels connected together, which increases the power output further.

A lot of research is being channelled into the development of new materials and methods to increase the efficiency of PV units. The most promising area of research is Organic Solar Cells (also referred to as plastic solar cells). These offer easier and cheaper manufacturing and more flexibility in the construction of solar panels.

The most exciting recent advances are the ability to print solar cells and panels, and the use of Perovskite. These will make PV cells increasingly cheaper and more efficient, but there are issues with durability. As with all forms of renewable energy there is an increasing impetus on research and development, but getting to large scale manufacturing and deployment is proving more difficult to achieve.

Hydro Power

This term applies to power derived from the energy of fast flowing or falling water.

Hydro power is one of the largest sources of energy, accounting for roughly 20% of the worldwide demand for electricity. Hydroelectric power is a preferred energy source in areas with heavy rainfall and hilly or mountainous regions. In countries where it is a naturally available resource it accounts for the majority of the energy supply.

Throughout history, the power of flowing water has been used to provide energy for mechanical devices that were used for a variety of tasks including flour and other mills, early manufacturing, and pumps for irrigation purposes. Although most of these methods were later powered by steam and then electricity or internal combustion engines, the development of turbines led to the introduction of water powered electricity generation, or hydroelectric power.

A conventional dammed hydroelectric generator is the most common source, but 'run of the river' installations are in use that use a smaller dam, or do away with it completely. There are also pumped storage facilities where, at low demand periods, water is pumped from a low level reservoir to a higher one, where it is stored until needed to produce more electricity at peak times.

Water from the reservoir flows through the intake and drives the turbine before flowing out at the bottom into the river.

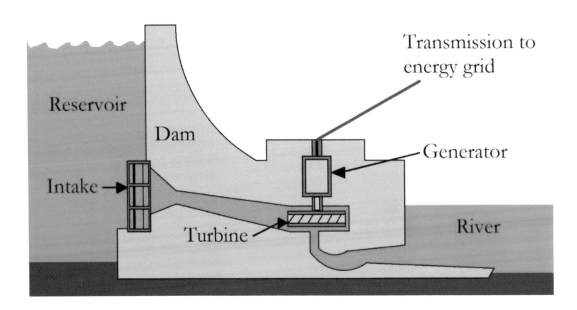

In an order that depends on where you live, the most famous hydroelectric dams in the world are: the Three Gorges Dam on the Yangtze River in China, the Aswan Dam on the Nile River in Egypt, and the Hoover Dam on the Colorado River which borders the U.S.A. states of Nevada and Arizona. There are many other hydroelectric dams around the world.

China is by far the biggest generator of hydroelectricity. Brazil, Canada and the U.S.A. are a close match for second place, with Russia, Norway, India and South America a bit further down the list.

Things change a bit when you look at how much of the total electricity used in a country comes from hydroelectric sources. Norway generates 99% of the country's electricity this way and there is room for further development. There are plans in place for Norway to export electricity to neighbouring countries. (So if you live in Norway, give yourself a pat on the back for being so green and go and celebrate by riding a bike or buying a BEV or FCEV).

Other countries that generate more than 95% of their electricity using hydro power are (in no particular order) Zambia, Albania, Belize, Paraguay, Bhutan, Burma, Mozambique, Ethiopia, Lesotho and Iceland.

Many of the negative environmental impacts of hydroelectric power come from the associated dams, which can interrupt the migrations of spawning fish such as salmon, and permanently submerge or displace ecological and human communities as the reservoirs fill. In addition, hydroelectric dams are vulnerable to water scarcity. In August 2021 Oroville Dam, one of the largest hydroelectric power plants in California, was forced to shut down due to historic drought conditions in the region.

Energy from our seas and oceans

<u>Wave Power</u>

This is a method of harvesting the energy of waves in water (usually a sea or ocean) and putting that energy to useful work. This can be water desalination, powering pumps (to pump water into reservoirs for example), a few other applications and, most importantly, electricity generation. A machine that can make use of wave power is generally known as a Wave Energy Converter (WEC). Let's take a look at the four most common methods of extracting energy from waves currently being used.

Point Absorber Buoy

These devices are the most common form of what are termed Oscillating Wave Surge Converters. These devices float on the surface of the water, held in place by cables attached to the sea floor. The buoys use the rise and fall of the surface swells to drive hydraulic pumps that generate electricity.

Surface Attenuator

Similar to Point Absorber Buoys but with multiple floating segments connected to each other. The swells create a flexing motion that drives hydraulic pumps to generate electricity.

Oscillating Water Column

The swells compress air in an integrated chamber that forces air through a turbine to create electricity. The process of the air being forced through the turbines creates quite a lot of noise, making these the noisiest method of converting wave power.

Overtopping Device

These are long structures that use wave velocity to fill a reservoir to a higher level than the surrounding ocean. The greater height of the reservoir is used to run water through low-head turbines that generate electricity.

The European Marine Energy Centre (EMEC) Ltd was established in 2003. Based in Orkney in the U.K. it was the first centre of its kind in the world to provide test sites and facilities for developers of both wave and tidal energy. Other research facilities now include the marine energy department of the National Renewable Energy Laboratory (NREL), which is a national laboratory of the U.S. Department of Energy.

Wave Power is still regarded as an emerging technology but is has the potential to become an important part of the renewable energy sector. There are some indications that wave power installations could reduce coastal erosion and help create a stable marine environment along the coast.

Tidal Energy

This exploits the energy produced by the tidal waters. The power of these fast-flowing sea currents is often magnified by narrow channels in the sea bed, headlands, inlets and straits.

Tidal Turbines

These are very similar to wind turbines. Instead of wind, the flowing current turns the turbine blades. They are able to generate more electricity than wind turbines because the water is denser than the air. Unfortunately this also means that tidal turbines need to be much stronger and are thus more expensive to manufacture and install.

There are two types of tidal turbine, horizontal axis and vertical axis. Horizontal axis is the same configuration as a wind turbine whereas vertical axis has the blades rotating sideways around the central pillar, a bit like a turnstile, or revolving door.

Horizontal axis turbines are arranged in similar patterns to wind farms, or used in a tidal barrage, which are low walled dams, usually installed at tidal inlets or estuaries. They are similar to hydroelectric dams, with turbines located along the bottom of the barrage being turned by the incoming and outgoing tides. Tidal barrages are the most effective way to harness tidal energy but are also the most expensive and have a larger ecological impact on the local ecosystem than a farm or tidal fence.

Vertical axis turbines are usually arranged into a tidal fence and installed in inlets and fast moving streams. They do not require the large concrete construction involved with barrages and have less impact on the surrounding ecosystem.

The first large scale tidal power system is the La Rance Tidal Power Station in Brittany, France. It is a tidal barrage with 24 turbines and opened in 1966. It currently generates enough electricity to power around 50,000 residential houses per year.

In 1968 the Kislaya Gubla station was opened in Russia, followed by Jiangxia in China 12 years later. The Annapolis Royal Generating station opened in Canada in 1984 but was decommissioned in 2019 due to environmental concerns and the failure of a crucial component.

In the U.K., Strangford Lough Sea Gen opened in 2008 and was decommissioned in 2019.

More recently, South Korea opened the Uldolmok tidal power station in 2009, followed two years later by the world's largest tidal power station at Sihwa Lake.

The Netherlands opened the Eastern Scheldt Barrier tidal power plant in 2015 and the U.K.'s Bluemell Sounds tidal stream array opened in 2016, followed by MyGen in the Penland Firth in 2018.

Russia, India, South Korea, Canada and the U.K. all have proposals for further tidal power generators at various stages of development. The U.S.A. currently has no tidal power stations but there are a few in the development stages.

The main drawbacks and considerations that need to be taken into account with tidal power are the early stage development costs, disruption to marine life, unpredictable weather, corrosion of the equipment and maintenance costs, silt formation (which leads to algae blooms that reduce water quality) and the fact that the tides only run strong for around ten hours per day, meaning they run at a lower capacity for at least 14 hours per day.

Geothermal

Thermal energy is a natural process that generates heat and stores it near or on the surface of the Earth. The differences in temperature throughout the molten core of the planet carries heat towards the surface, where it heats the mantle (the covering of rock that forms the surface of the Earth). Geothermal energy production uses the energy stored by this process.

At a basic level, geothermal energy has been used by mankind since Palaeolithic times, through hot springs being used for cooking, bathing and heating. The earliest recorded use is by the Paleo Indian culture in the Americas dating back nearly 10,000 years. The Romans also used it for heating buildings. Recently it has been harnessed to create geothermal electricity. The original geothermal power plants are called Dry Stream and they use steam generated by a hydrothermal reservoir directly to spin the turbines.

The most commonly used method is now Flash Steam. This utilises highly pressurised reservoirs where the water temperature exceeds $180°$ C. The pressure pushes the superheated water into a lower pressure tank on or near the surface. The lower pressure causes some of the water to be vaporised (flashed) into steam that turns a turbine. Any water remaining is injected back into the hydrothermal reservoir to be used again. This method is more efficient than Dry Stream.

The newest technology is called Binary Cycle whereby the hydrothermal reservoir water is used to heat a liquid with a lower boiling point. The hot water is cycled back to the reservoir and the working fluid vaporises and turns the turbine. This method accounts for most of the recent growth in geothermal energy generation globally. There is virtually no pollution created by these modern systems.

Countries that have natural hot springs and geysers, or are volcanically active, can make use of this resource relatively easily. This includes parts of the U.S.A., Iceland, China and Japan. More recently, other areas have been opened up by new technologies. Geothermal energy is considered to be cost effective, reliable, sustainable and environmentally friendly.

Biomass

Biomass is material from living or recently living organisms. In energy terms it often refers to any plant based material but it can also mean animal or vegetable derived materials. Energy generated using biomass techniques is called 'Bioenergy'.

The material that is captured in fossil fuels is fossilised biomass, but using it returns CO2 into the atmosphere that was captured millions of years ago. Using biomass returns CO2 that has been absorbed very recently and is part of what is called 'The Carbon Cycle'.

'What's that? Some kind of expensive bicycle?' I hear you ask. No, of course not. It is basically the cycle of life on Earth. A series of processes combine to make carbon compounds that are released into the environment. Carbon Dioxide in the atmosphere is converted into living tissue by the process of photosynthesis. The CO2 is returned to the atmosphere by respiration (breathing), the decay of dead organisms, and burning biomass.

There are five basic categories of material that can be used to generate energy. These are:

Virgin Wood from forestry, wood processing and arboreal activities.

Energy Crops – high yield crops grown specifically for energy applications.

Agricultural Residue which is the organic residue left after harvesting and processing.

Food Waste – food and drink manufacturing and consumer waste.

Industrial Waste – from manufacturing and industrial processes.

Biomass can be used to generate heat and electricity, liquid biofuel or combustible biogas. This is achieved by means of two methods; Thermal Conversion and Chemical Conversion.

Thermal Conversion

As the name suggests, this involves using heat to convert biomass into a different chemical form. The main processes are:

Combustion

Yes, setting fire to it is the simplest method of converting biomass to energy for use in space heating, heating a liquid (normally water) for central heating, and other similar uses. It can also be used to generate steam for electricity generation.

Gasification

The same process described in the section about producing, processing and storing hydrogen.

Pyrolysis

This is the thermochemical decomposition of organic material in the absence of oxygen. It produces gases and liquids, leaving behind a carbon rich solid residue. Pyrolysis is the precursor to gasification and is used to produce charcoal, methanol, PVC, syngas and loads more. It also plays an important role in cooking – specifically baking, frying, grilling and caramelizing.

Chemical Conversion

So, this will be using chemical reactions instead of heat, then. The main process is Biochemical Conversion, which uses bacteria or other micro-organisms that produce enzymes to break down the biomass. The three most used methods for this are as follows;

Anaerobic Digestion (AD)

Bacteria break down organic material in the absence of air. This produces three things - biogas that contains methane, a solid residue similar to compost, and a liquor that can be used as a fertiliser. It has been used to process sewage since the 19th century.

Fermentation

This is the same process used in brewing and wine making. It converts sugars into alcohol and can be further distilled to obtain bio-methanol, which can be used as a fuel or added to petrol.

Composting

Similar to AD, this process uses different bacteria. As any keen gardener will know, it also generates a lot of heat, which can be used to generate power through a heat pump.

Bioenergy currently generates around 100 gigawatts of power globally and provides around five percent of world transport fuel. You will use one of the by-products the next time you have a charcoal barbecue and it's already in all transport fuels including aviation. Demand for biofuels increased globally by around 6% in 2022.

Waste and wood biomass use for generating electricity is growing every year.

Nuclear Power

Nuclear energy is a form of energy emitted by the nuclei of atoms, made up of protons and neutrons. This source of energy can be produced in two ways: fission - when the nuclei of atoms split into several parts - or fusion - when the nuclei fuse together.

The nuclear energy that is being harnessed around the world today to produce electricity is through nuclear fission, while the technology to generate electricity from fusion is still in the research and development phase.

Although the understanding and use of radioactivity only started around 130 years ago, it has advanced quickly and become a hotly debated subject, mostly due to the development of nuclear weapons.

The element Uranium was discovered in 1789 by a German chemist called Martin Klaproth. He named it after the planet Uranus. Then nothing much happened until ionising radiation was discovered by Wilhelm Rontgen in 1895 when he was able to generate X-rays. Other types of ray were soon identified, including alpha, beta and gamma. In 1896 Pierre and Marie Curie gave the name radioactivity to the phenomena. They went on to isolate the elements radium and polonium. Radium was later used in medical treatments.

In 1898 Samuel Prescott showed that radiation destroys bacteria in food.

The science of atomic radiation, atomic exchange and nuclear fission continued to be studied, but from 1939 to 1945 most development was focused on the atomic bomb.

After World War II the focus switched to controlling and harnessing the energy for use in naval propulsion, and generating electricity.

The first nuclear reactor to produce electricity (initially a very small amount) was the small Experimental Breeder reactor (EBR-1) which started up in Idaho, U.S.A. in December 1951.

There were nuclear development programmes being run by various governments around the world using slightly different techniques.

Since the process was first commercialised in the 1970's, about 65 billion tonnes of carbon dioxide that would have been produced using fossil fuels to generate the heat has been saved. It is also used to power large submarines.

Unfortunately, due to concerns about radioactive waste disposal, its possible use to make atomic weapons, and nuclear accidents such as Three Mile Island, Chernobyl and more recently Fukushima, there has been a steady increase of opposition to Nuclear Power.

Modern reactors are much safer and more efficient than their predecessors and produce less waste. In 2012 the World Nuclear Association reported that nuclear electricity generation was at its lowest level since 1999. Nuclear power now provides about 10% of the world's electricity with around 440 power reactors in use globally.

Currently there is a growing interest in Small Modular Reactors (SMRs), which may well be the next boom industry in renewable energy. SMRs are small, low cost reactors and micro-reactors that are assembled in factories and then shipped and installed on site. SMRs offer large savings in cost and construction time. They will usually cover an area equal to one or two football fields and are ideal for small rural and isolated communities, as well as providing local power and additional grid production.

Compared with current large scale nuclear reactors, SMR designs are generally simpler and safer. They have lower fuel requirements, running between 3 and 7 years from each refuelling compared to 1 to 2 years for large scale reactors. The latest designs could even run for decades.

Russia's Akademik Lomonosov, the world's first floating nuclear power plant that began commercial operation in May 2020, is producing energy from two 35 MW(e) SMRs. Many companies, including Rolls Royce, are taking orders from governments around the world and the first ones should be operating by the end of 2023. You will be hearing a lot more about SMRs in the coming years.

Of course, we are talking about the process of nuclear fission. There is much research being done into nuclear fusion, which is the process that takes place inside the sun. When two atoms of a light element such as hydrogen are subjected to extreme high temperature and pressure they combine and form a single heavier element such as helium, releasing massive amounts of energy in the process, much more than fission. If this can be controlled and harnessed it will be one of the biggest breakthroughs in history and has been described as the 'holy grail' of energy production.

In December 2022 the National Ignition Facility at the Lawrence Livermore National Laboratory (LLNL) in California announced that they had overcome one of the major barriers and produced more energy from a fusion experiment than was put in. They used a 192 beam laser array to heat a peppercorn sized capsule containing hydrogen to 100 million degrees Celsius (hotter than the centre of the Sun) and 100 billion times the normal Earth sea level air pressure. They used 2.05 megajoules (MJ) of energy and generated 3.15 MJ of output, which is enough power to boil fifteen to twenty kettles.

A lot of investment is being made globally into the various research projects and there are currently three main designs.

Magnetic confinement fusion (MCF). This generates magnetic fields to confine plasma in a donut shaped vessel (torus).

Inertial confinement fusion (ICF). High power lasers are used to compress and heat the fuel.

Magnetised target fusion (MTF). Sometimes also referred to as magnetised inertial fusion (MIF), this combines MCF and ICF using higher density plasma and lower power lasers.

Useable, scalable nuclear fusion reactors are still unlikely to become a reality before 2040.

Wind Power

Harnessing the power of the wind has played a pivotal role in human history. Sailing - using wind propulsion to move craft over water, ice, or land, has been in use since at least 5,000 BC, when the first records of its use show sailing boats on the Nile River. The earliest representation of a sail ship is on a painted disc found in Kuwait, estimated to date from between 5,500 and 5,000 BC. All major civilisations since then have used it to explore, trade, and supply their empires.

From the middle ages there was a steady progress in sailing technology, ship building, and navigation This led to ever increasing exploration, the development of military and merchant navies, advances in naval warfare, and vast amounts of trade. It is still very much in use today and has a large range of participants. Probably the best known sailing race is the Americas Cup.

Windmills are another early adoption of wind power. According to historical data, windmills were first used in China to pump water from around 2,000 BC. The Persians used them to pump water and mill grains from around 500 AD.

In Europe, windmills used to crush grain began appearing from around 1100 AD. In the 1300's the Dutch used windmills to drain water from low-lying regions. With the invention of steel blades, windmill use exploded in the late 1890's for pumping water and generating electricity. They were used extensively in the settling of the U.S.A. and Australia.

A lot of the electricity generation and pumping was taken over by steam and internal combustion engines and, more recently, gas turbines, but the use of wind power has remained quite strong. Interest peaked during the oil crisis in the early 1970's, along with a more general interest in renewable and alternative energy sources. Wind turbines have been continuously developed and improved since then, assisted by various government incentives (including tax credits) in different countries around the world for renewable energy production.

Modern wind turbines are complicated pieces of machinery. There are currently two basic types of wind turbine – vertical-axis, most often associated with the Darrieus model, which resembles a whisk or eggbeater, and the more commonly used horizontal-axis, which include the traditional windmills and modern three blade wind turbines. Groups (or arrays) of wind turbines are called wind farms and this is the most widespread usage of them. Wind farms can be on or off shore and generate electricity that is put into energy grids for large scale consumer purposes. They are also now being used to generate green hydrogen.

The main components of a horizontal-axis wind turbine are;

- A tower that supports the rest of the structure;
- A drive train that includes a gearbox and generator;
- The blades (or rotors) that convert the wind energy into rotational energy to drive the generator;
- Other equipment including controllers for adjusting the blade pitch and direction to keep them facing the wind and stop them from spinning too fast, pumps and lubrication, electrical cables, ground support and interconnection facilities.

Horizontal-axis wind turbines come in a variety of sizes and power ratings. A small individual use turbine stands around 30 feet high with rotors between 8 and 25 feet in diameter, while the largest ones (used in offshore facilities) are over 75 storeys high with rotors up to around 700 feet in diameter (approximately the length of two football pitches).

The world's largest wind turbine is currently the General Electric Haliade-X in Rotterdam. It was first tested in October 2019 and then upgraded to 13-MW. It set a new 24-hour power generation record in January 2020 with an output of 288 MW/h in one day. The Haliade-X stands 260 m (853-ft) high and has a rotor diameter of 220 m (721-ft). The rotor blades are 107 m (351-ft) long (which is twice the wingspan of the Angel of the North) with a 6 m (19-ft 8-in) hub. One single rotation of the rotor produces enough electricity to power an average home for at least two days.

In China, the MingYang Company has a prototype of an even larger wind turbine ready to be installed in 2023 and be commercially productive in 2024. It will be 242 m (794-ft) tall, rated at 16 mw and is expected to generate enough electricity to power 20,000 homes over its 25 year service life.

93.6 GW of new wind power capacity was added worldwide in 2021 and the total wind capacity is now over 837GW. China had the most new onshore installations, followed by the U.S.A., Brazil, Vietnam and the U.K. For offshore installations China is also leading the way, followed by the U.K., Denmark, Vietnam and the Netherlands.

Vertical-axis wind turbines are gradually gaining popularity as they are smaller, easier to install and omni-directional, which means that they don't need to be rotated to face the wind direction. Vertical-axis turbines can operate in lower wind conditions and can cope with extreme conditions more easily. They have fewer moving parts, are more suitable for urban, suburban and even individual use and can be easier to maintain, but they are not currently as efficient as horizontal-axis turbines.

Recommended further reading

What is Concentrated Solar Power? (solarguide.co.uk)

How Solar Panels Work - A Guide For Dummies - Go Greena Blog

Hydroelectric Power (therenewableenergycentre.co.uk)

Marine Energy : EMEC: European Marine Energy Centre

Bioenergy - Fuels & Technologies - IEA

Nuclear Power Today | Nuclear Energy - World Nuclear Association (world-nuclear.org)

Wind Energy Timeline – From Persian Windmills Crushing Grains to Vesta's Wind Turbines Churning out 8 MW of Output | AltEnergyMag

Advantages and disadvantages

Large scale energy production (for power grids)

Solar Towers

These require desert or semi desert conditions and will, therefore, require the electricity generated to be transmitted over quite large distances. This means that some of the power is lost before it gets into the grid. A major advantage in their location is that the land is very difficult to use for any other purpose, and there are no large local populations that will be adversely affected by their construction and use.

Because this is a relatively new technology that is still being developed and requires large amounts of land and large structures, it is expensive to set up.

Although it depends on long periods of strong direct sunlight, this method continues to generate power for a while after the sun has set due to the inbuilt heat energy storage capacity. Once it has been running for a short period of time the electricity production costs are quite low, especially when combined with the latest storage batteries.

So, for the regions of the world where the conditions are right, this is a practical long term technology.

Nuclear Power

These are, quite frankly, hideously expensive and complicated to set up. Once they are up and running, however, the modern facilities are safe and efficient. The main drawbacks are disposing of the nuclear waste, and the fact that they have to be situated away from any major settlements due to the strict regulations around radioactivity contamination risks.

They generate power continuously and can be adjusted to produce varying amounts depending on demand. This makes them an attractive option for major grid networks.

The upcoming roll out of Small Modular Reactors is going to transform the market and add more local electricity generation to support the grid.

Wind Power

The very large size of the individual turbines makes them expensive to build and install. They also require large areas of land or sea. Off shore wind farms have similar problems to solar towers in that the electricity generated needs to be transmitted over long distances to be transferred into the grid. This is less of a problem for onshore facilities, and the ground they occupy can be used for other purposes, most notably agriculture.

The power supply is variable and, as each turbine requires power to keep the various lubrication and adjustment equipment operating, there is some debate about their overall efficiency. Some research seems to indicate that the nett power production is much lower than generally reported, although this is hotly disputed. Another possible disadvantage is that they have a limited lifespan (again there is a lot of debate around the actual figures) and so will need to be replaced at regular intervals. The latest figures indicate a lifespan of around 25 years, with a gradual decline in efficiency that can be mitigated by replacing worn out or damaged parts. Wind turbines installed in the early 1990's currently produce on average about 75% of their original output. This means that the amount of pollution created in constructing and maintaining the wind farms is unlikely to decrease and the costs will continue to be an important factor.

There are also concerns about the impact on the local environment based around the low resonance vibrations they produce in the ground and the number of local and migratory birds that fly into the turbines.

It is also apparent that they create turbulence and thus changes to the local wind patterns behind them. This has been measured in associated local temperature and weather pattern changes. If you subscribe to the 'Butterfly Effect' theory then the large scale use of these farms has to have an effect on the global weather patterns which will only increase as more are built.

One other problem is that this is an industry with huge amounts of investment and research. Some of the advances (bladeless technologies being a prime example) will make the current turbine designs obsolete before they reach the end of their useful lifespan. This poses the question of whether to hold back the new technologies until the current turbines need to be replaced. The impact this will have on the companies that manufacture and install the equipment also has to be considered, especially when you factor in the huge subsidies currently paid by governments to support the industry.

Hydro-electric

Where the landscape and local environment are suitable, this is an obvious choice. Constructing dams is quite expensive but this is an ancient technology and very well understood. Once constructed they are reliable and relatively cheap to maintain.

The impact on the local environment can be severe if large reservoirs are required, and changing the flow of major rivers can adversely affect the environment for hundreds of miles downstream.

Overall, if environmental impacts are kept to a minimum, this is a safe, reliable and relatively cheap way of manufacturing electricity on a medium to large scale.

Geothermal

In countries where it is readily available this is a cheap and reliable source of energy. As well as using it to generate electricity, it can be used for heating and, of course, natural hot springs for relaxing, bathing and cleansing.

Tidal

Still in its infancy, tidal energy has plenty of potential but is expensive to set up.

Whilst tidal energy is predictable and easier to harness than wind power, it uses a similar technology of turbines. It is very expensive to install and maintain but after that the cost of generating electricity is very low. At the moment it is limited to near-shore locations and there are concerns about the impact on the local environment.

Wave

This is a relatively simple and cost effective way of generating electricity. It is ideal for near shore installations and has a minimal impact on the marine environment. In fact, some research shows that a 'wave farm' could create a more stable coastal habitat and reduce coastal erosion.

There was a lot of promising research done into wave power in the 1970's but it never got much backing and only recently has it begun to be taken seriously. It is in a similar position to tidal energy at the moment but has the advantage of being able to generate electricity continuously with much cheaper installation and running costs. This is a technology that really needs to be taken more seriously.

Solar Arrays

Solar cell efficiency is increasing all the time, whilst the size and cost is reducing at a similar rate. For large scale use it requires a lot of land, but, as with wind farms, the land can still be used for farming and livestock.

Solar arrays are relatively inexpensive to set up and have minimal running costs. As with all forms of solar energy, they only generate electricity during the day.

Small scale energy production (for local and individual use)

Fuel cells

For vehicles, this is a very promising power source to replace diesel and petrol internal combustion engines. There is a growing interest in fuel cells globally and they are becoming cheaper and more efficient all the time. For them to really become mainstream, all that is required is a fuel network, and this is starting to be addressed.

There are a few major advantages that FCEVs have over battery electric vehicles. They currently have a greater range and only take five minutes to refuel, whereas the best of the BEVs still require around eight hours to fully recharge, and while this is reducing all the time, it will almost certainly be at least ten years before the recharging time is less than an hour. Advances in storage batteries is extending their range as well, but the electricity to charge them comes from the grid. As most countries do not have much spare capacity, the mass use of BEVs will cause major problems without large scale investment to increase the grid capacity. How the electricity is generated directly affects how green the use of BEVs can be.

FCEVs provide a very similar user experience to current vehicles in that you drive them until they need fuel and you then fill it up in a few minutes and continue on your way. They have the added advantage of being a mobile electricity generator.

Fuel cells can also be installed in houses and small buildings to supply electricity on demand. They are proving to be a cheap, reliable and clean source of electricity.

This could be the most important renewable technology for the future and needs to be promoted and supported as much as possible.

Solar Panels

There has already been a lot of take up globally for individual houses and small buildings and as the technology is constantly advancing this should only increase. It is ideal for remote areas. A start-up business in Kenya called AzuriTV has been offering a solar powered satellite TV system designed for people in remote rural locations that is really helping to improve the lives of people, especially in remote farming communities.

This is another important renewable technology that needs as much support as possible.

Biomass

This is a great way to provide electricity in a local community environment. It uses waste materials (and can be installed as part of a water purification process), provides useful by-products and is a clean, reliable and relatively cheap way of producing electricity on a low to medium scale.

Wind

This is expensive and unreliable. My daughter's school had a small turbine installed in the mid 2000's and all of the pupils were very disappointed when the gauge showing the amount of electricity produced rarely rose enough to power one 60 watt lightbulb. As far as I can tell, wind turbines are not generally considered suitable for small scale use, although some modern building designs incorporate an Archimedes screw type of arrangement to capture wind at the edges of the building. Vertical-axis turbines are much more suitable for small scale generation.

Geothermal

In areas where it is naturally available, this is an ideal source of heat and power for individual to medium scale use and is cheap and reliable.

Tidal

This is expensive to set up and maintain and can only generate electricity at certain times of day. It is not suitable for individual or small scale power supply.

Wave

This appears to be a promising method of producing electricity for seaside towns and communities and once set up should be cheap and reliable. It could also improve the local marine environment.

Hydroelectric

As with wave and geothermal sources, this is a reliable way to generate electricity on a small to medium scale where it is naturally available.

Combining different technologies

It is obvious to me that the best way forward with green and renewable energy is to combine the different technologies.

For grid networks, more needs to be done to use each type where it is naturally available. Wind farms already receive huge amounts of investment and subsidies. We need to commit more resources into developing the various ways of generating power from the oceans, seas, rivers and waterways, particularly wave and tidal power. By combining all of the natural resources available, increasing the use of biomass generation, and investing in new technologies such as small modular nuclear reactors, I am convinced that most countries could do away with the need for coal, nuclear and gas powered generators within a relatively short period (between 25 and 50 years, depending on the country and the amount of external investment). By encouraging the take up of small scale and individual power generation, the demands on the grid would be reduced as well. There is also a method called combined heat and power (CHP), or cogeneration. It basically involves using the excess heat produced by any method of energy generation to supply hot water and increase the efficiency of the generating process.

For transport, housing and factories there are a few interesting options. The most promising and potentially viable development combines solar power, hydrogen fuel cells and storage batteries. Solar panels generate electricity which can be used directly and keep the storage batteries topped up. Excess electricity is used to generate hydrogen fuel by electrolysis, which is used in fuel cells to provide electricity on demand. The hydrogen can also be used as an alternative to natural gas for cooking and heating. There is a lot of research and development going into this and early adopter systems are already available.

I would like to see a lot more joined up thinking instead of promoting one way of generating renewable power over the others. This is beginning to happen, with some of the electricity generated by wind and solar power being used to power hydrogen production.

Other combinations aimed at cleaning up the environment are beginning to surface. I would like to see wave power installed around the coast combined with shellfish and crustacean farming and 'rewilding' using sea grass and other naturally occurring marine vegetation. The wave power generators should reduce coastal erosion and help create a safe marine environment, as well as powering coastal towns and villages. The shellfish (carefully selected for the local environment) provide an income and fresh food supply whilst filtering and cleaning the water. Most marine life reproduces near the shoreline so better, safer conditions would improve the breeding conditions and help reverse the decline in numbers. More fish and crustaceans could be caught locally and should improve in quality as well.

Everyone needs to start having proper discussions about what projects need to be supported, from local communities right the way up to top ranking politicians and business leaders. If we all work together and support each other we can have cheap, reliable and clean power supplies which will benefit the whole planet.

Nearly finished

The rest is up to you

Everyone needs to be more involved with the huge task of changing from fossil fuels to renewable and less polluting alternatives as quickly as possible for the health of the planet and future generations. Whether this is actively promoting alternative energy, joining and supporting pressure groups, working within the industry, investing in green technology, or just telling people about it, you can make a difference.

There is one last thing that has occurred to me and that is the whole question about global warming and climate change. There are strong arguments on all sides as to whether it is happening and the amount of influence that humankind has on it. I think that it is a major mistake to present it in this fashion. The world has heated up and cooled down constantly since the planet first formed, sea levels have risen and fallen dramatically and researchers are finding more and more evidence of these events all the time. The one thing that no-one can successfully argue against is that pollution is bad for everyone and everything. This is where the emphasis needs to be placed to stop the endless arguing that seems to be the mainstay of everything these days. We need to stop accepting all the negativity and concentrate on being positive and proactive.

I hope that you have enjoyed this book and now feel confident enough to persuade everyone you know to take positive action in order to reduce our impact on the environment and save the world for future generations before it's too late.

Recommended further reading

Home | Bill Gates (gatesnotes.com)

New Atlas - New Technology & Science News

Renewable Energy | Virginia Cooperative Extension | Virginia Tech (vt.edu)

gasworld - the latest industrial gas news, conferences, directory and much more

Linde - Making our world more productive

Reaction Engines - Making Beyond Possible

Thank you for reading my book. I hope that you found it interesting, informative and useful. As a writer who is always striving for perfection, the views of my readers are very important. Please take a few moments to rate and review my book. All feedback is greatly appreciated and your review will also be useful to other people who might be interested in my books.

References

Introduction

Two Superpowers We Wish We Had | Bill Gates (gatesnotes.com)

Climate change: surely the most important news story of our age? | Membership | The Guardian

(6) It's Time for an Optimistic Conversation About Climate Change | LinkedIn

IPCC — Intergovernmental Panel on Climate Change   www.ipcc.ch

The World Economic Forum (weforum.org)   www.weforum.org

The Earthshot Prize: Urgent Optimism & Action to Repair the Planet   www.earthshotprize.org

ClimateChange | United Nations   www.un.org/climatechange

Electricity

Turbine - Energy Education   www.energyeducation.ca

Electric power transmission - Wikipedia

A Timeline Of History Of Electricity (electricityforum.com)

Atom: Definition, Structure & Parts with Labeled Diagram (sciencefacts.net)

Electricity transmission and distribution | Institution of Civil Engineers (ICE)

Voltaic Pile – 1800 - Magnet Academy (nationalmaglab.org)

How do electric motors work? - Explain that Stuff

Batteries

The Evolution of Battery Technology (visualcapitalist.com)

LION Smart - The History of Battery Technology: Evolution of Energy Storage

Electric vehicles guide - EV charging, models and costs UK 2023 (zap-map.com)

History of Electric Cars | Sytner Group

What are lithium batteries made of and what are their pros and cons? (power-technology.com)

Lithium-ion battery - Wikipedia

Why We Should All Love Hydrogen

What Is Hydrogen?

Hydrogen economy - Wikipedia

It's Elemental - The Element Hydrogen (jlab.org)

What is hydrogen? Atoms and Chemistry - Quatr.us Study Guides

Structure of the atom - Atomic structure - AQA - GCSE Combined Science Revision - AQA Trilogy - BBC Bitesize

The Element Hydrogen -- Hydrogen Atom (worldofmolecules.com)

Hydrogen Facts – 12 of the best from Chemicool.com

Hydrogen - Wikipedia

Hydrogen | Air Liquide Energies

Hydrogen in Human History

Timeline of hydrogen technologies - Wikipedia

1966 GM Electrovan - First Hydrogen Fuel Cell Vehicle | Hydrogen Cars Now

Hofmann voltameter - Wikipedia

How hydrogen is produced, processed and stored

Universal Industrial Gases, Inc...Hydrogen H2 Properties, Uses, Applications (uigi.com)

Alternative Fuels Data Center: Hydrogen Production and Distribution (energy.gov)

Hydrogen Production | Department of Energy

Breakthrough in hydrogen fuel production could revolutionize alternative energy market | VTx | Virginia Tech

Gasification Systems | Department of Energy

How Gasification Works | HowStuffWorks

Hydrogen Production: Electrolysis | Department of Energy

How to Make Hydrogen Gas (4 Methods) (thoughtco.com)

Hydrogen Production: Photobiological | Department of Energy

photocatalytic-water-splitting | Solar Energy & Advanced Materials Research Group - UCL – University College London

Biohydrogen - Wikipedia

Biological hydrogen production - University of Birmingham

Hydrogen Storage | Department of Energy

Hydrides - Chemistry LibreTexts

Hydride | chemical compound | Britannica

First of its Kind Discovery in Mali: Vast Reservoirs of Clean Hydrogen Gas - Hydrogen Central (hydrogen-central.com)

Helios Aragon | Solutions to the green hydrogen deficit. (helios-aragon.com)

Practical application and uses of hydrogen

Uses of Hydrogen - Uses Of

The Centre for Fuel Cell and Hydrogen Research, Chemical Engineering, University of Birmingham - University of Birmingham

What is Hydrogenation? (with pictures) (delightedcooking.com)

How Fuel Cells Work | HowStuffWorks

Fuel cell - Wikipedia

Fuel cells | Johnson Matthey

FCF_Smart.pdf (pemfc.de)

What are Microbial Fuel Cells? - How do fuel cells work, Info on Microbial Fuel Cells (altenergy.org)

Hydrogen Power | The Water Page

Types of Fuel Cells | Department of Energy

Where are we now?

Hydrogen Products - Pure Energy Centre

Riversimple launches Rasa, a hydrogen-powered city car for the masses (newatlas.com)

A Hydrogen Car Just Set The Record For The World's Longest Continuous Journey : ScienceAlert

Hydrogen, hydrogen everywhere... - BBC News

Fuel Cells Works - Fuel Cells, Energy And Hydrogen News

ITM POWER (itm-power.com)

Fuel cell vehicle - Wikipedia

Hydrogen (FCEV, Hydrogen-powered Engine) | Beyond Zero | Mobility | Toyota Motor Corporation Official Global Website

Hydrogen refuelling project enters second phase - AirQualityNews

Norway and South Korea could team to support hydrogen fuel cars (hydrogenfuelnews.com)